GRAIN BOUNDARY DIFFUSION AND PROPERTIES OF NANOSTRUCTURED MATERIALS

GRAIN BOUNDARY DIFFUSION AND PROPERTIES OF NANOSTRUCTURED MATERIALS

Yu.R. Kolobov, R.Z. Valiev, G.P. Grabovetskaya,
A.P. Zhilyaev, E.F. Dudarev, K.V. Ivanov,
M.B. Ivanov, O.A. Kashin, E.V. Naidenkin

Viva Books

New Delhi | Mumbai | Chennai | Kolkata | Bangalore | Hyderabad | Kochi

First Indian Edition 2008

VIVA BOOKS PRIVATE LIMITED

- 4737/23 Ansari Road, Daryaganj, New Delhi 110 002
  E-mail: vivadelhi@vivagroupindia.net, Tel. 42242200
- Plot No. 76, Service Industries, Shirvane, Sector 1, Nerul, Navi Mumbai 400 706
  E-mail: vivamumbai@vivagroupindia.net, Tel. 27721273, 27721274
- Jamals Fazal Chambers, 26 Greams Road, Chennai 600 006
  E-mail: vivachennai@vivagroupindia.net, Tel. 28290304, 28294241
- B-103, Jindal Towers, 21/1A/3 Darga Road, Kolkata 700 017
  E-mail: vivakolkata@vivagroupindia.net, Tel. 22836381, 22816713
- 7, Sovereign Park Apartments, 56-58, K. R. Road, Basavanagudi, Bangalore 560 004
  E-mail: vivabangalore@vivagroupindia.net, Tel. 26607409, 26607410
- 101-102 Mughal Marc Apartments, 3-4-637 to 641, Narayanguda, Hyderabad 500 029
  E-mail: vivahyderabad@vivagroupindia.net, Tel. 27564481, 27564482
- Beevi Towers, First Floor, SRM Road, Kaloor, Kochi 682 018, Kerala
  E-mail: vivakochi@vivagroupindia.net, Tel: 0484-2403055, 2403056

www.vivagroupindia.com

Published by arrangement with
Cambridge International Science Publishing
7, Meadow Walk, Great Abington
Cambridge CB1 6AZ, UK

ISBN-10: 81-309-0832-8
ISBN-13: 978-81-309-0832-8

Published by Vinod Vasishtha for Viva Books Private Limited, 4737/23, Ansari Road, New Delhi-110 002. Printed and bound by Anand Sons, Delhi-110 092.

Contents

Introduction

Grain boundary diffusion is one of the main physical processes which determine the development of grain boundary plastic deformation, intergranular failure, recrystallisation and coalescence of grain-boundary phases in polycrystals of metals and alloys at elevated and high temperatures, including in the conditions of the effect of diffusion flows of impurities from an external medium (coating) [1].

In the group of the latest achievements in the area of the theory of grain boundary diffusion, it is especially important to mention the general classification of the conditions of grain boundary diffusion for stationary and migrating grain boundaries [3, 4], developed on the basis of the improved classic Fischer model [2].

Analysis of the physical meaning of the realisation of a specific regime of grain boundary diffusion in combination with the results of theoretical and experimental investigations of the processes of diffusion-induced migration of grain boundaries, grain boundary sliding and the manifestation of superplasticity has made it possible to justify the physical considerations on the possibilities of the transfer of the grain boundaries to the activated non-equilibrium state in the conditions of the effect of directional grain-boundary diffusion flows of the impurity atoms from an external source [1]. These considerations represent a further development of the previously proposed concept of the possibility of formation of such a state in the conditions of interaction of the grain boundaries with lattice dislocations [5, 6].

With the development of a new group of materials (nanostructured metals and alloys, produced by different methods), the physics of grain boundaries and the relationships governing grain boundary diffusion in the ultrafine grain state have become the subject of interest of an increasing number of investigators throughout the world. The characteristic feature of a nanostructured material is the small size of the crystals and, consequently, the large length of the internal interfaces. The nanostructured materials are characterised by a unique structure and properties many of which are of interest for practice [7].

Several methods of production of these materials have been developed. The majority of these methods are based on compacting powders produced by different methods. In the group of these powders an important position is occupied by the ultrafine powders, produced by gas condensation in the atmosphere of an inert gas or by a plasma chemical method, aerosol and chemical synthesis, and also milling of powders in a ball mill, etc. Some of these methods have been used successfully for the production of bulk nanostructured materials. These materials are the subject of a large number of investigations of the special features of the nanocrystalline and nanostructured states. At the same time, there are a number of difficulties in the development of these methods of production of nanostructured materials. These difficulties are associated with the retention of some residual porosity in compacting, contamination of the specimens in the preparation and consolidation of the powders, and with the possibility of obtaining large geometrical dimensions of the produced specimens.

Many of these problems can be overcome using the methods of treatment by the effect of severe plastic deformation (SPD), i.e. plastic deformation in the conditions of high applied pressures [8, 9]. The task of the SPD methods is the formation of nanostructures in thick metallic specimens and blanks by refining the microstructure to the nanodimensions. It is well-known that by high deformation at low temperature, for example, as a result of cold drawing or rolling, it is possible to refine greatly the structure of the metals. However, the produced structures are, with a small number of exceptions, cellular structures or substructures having the boundaries with small-angle disorientation. At the same time, the nanostructures, produced under the effect of SPD in the optimum conditions, represent the ultrafine grained condition with mainly high-angle boundaries. In this case, many grains contain a specific substructure, associated with high distortion of the crystal lattice and, therefore, the dimensions of the crystals (the regions of coherent scattering, measured by x-ray methods) are usually smaller than 100 nm [8]. This also determines the formation of the nanostructured states in the SPD materials.

Under the effect of SPD it is possible to obtain very high deformations at relatively low temperatures in the conditions of high applied pressures. To realise these possibilities, special methods of mechanical the formation have been invented and developed: high pressure torsion, equal-channel angular pressing, etc. It has been shown that the bulk nanostructured specimens and blanks may be produced by the SPD methods from greatly differing metals and alloys, including many industrial alloys and intermetallics [8].

It is natural to assume that in the nanocrystalline ultrafine grained [7] and nanostructured materials [8]* processes of grain boundary diffusion play a significant role in the formation of the structure-sensitive properties at elevated temperatures. This relates to various diffusion-controlled processes, such as recrystallisation, low temperature plastic deformation in active loading and creep, superplastic flow, temperature-dependent internal friction and grain boundary deformation in fatigue.

In real polycrystalline materials, these processes control the relationships governing the development of plastic deformation and failure on different structural (scale) levels [10].

In the book, examination of the special features of the development of these processes in the nanostructured materials is preceded by the analysis of many relationships and models of grain boundary diffusion, including for nanocrystalline and nanostructured materials. The investigations of the diffusion permeability of these materials, carried out by Russian and foreign scientists, including the authors of the present book, have yielded unusual experimental results requiring clarification and indicating that the diffusion coefficients (mainly the coefficient of grain boundary diffusion) are several orders of magnitude higher than the appropriate values for the coarse-grained condition.

The monographs contains the main results obtained in the investigations of diffusion through the internal interfaces, which represent an important element of the structure of the nanocrystalline and the nanostructured materials. It is explained that the grain boundary diffusion determines the kinetics of many macroscopic processes. The data on the behaviour of materials with a high fraction of the grain boundaries are of considerable importance for all areas of the physics of polycrystalline materials, associated with the grain boundary processes.

The basis of the book is represented by the results, obtained by scientists at the Institute of Physics of Strength and Materials Science of the Siberian Division of the Russian Academy of Sciences and the Siberian Physical–Technical Institute under the supervision of Prof. Yu.P. Kolobov and the Institute of Physics and

---

*Here and in the rest of the book we use the terminology where the term 'nanocrystalline materials' refers to the nanomaterials, produced by the methods of powder metallurgy or electrodeposition, and the term 'nanostructured materials refers mainly to the alloys and metals produced by SPD.

Technology of Advanced materials at the Ufa State Aviation Technical University and of the supervision of Professor R.Z. Valiev.

Chapter 1 was written by A.P. Zhilyaev, R.Z. Valiev, chapter 2 – M.B. Ivanov, Yu.R. Kolobov, chapter 3 – Yu.R. Kolobov, M.B. Ivanov, chapter 4 – G.P. Grabovetskaya, Yu.R. Kolobov, E.F. Dudarev, K.V. Ivanov, A.P. Zhilyaev, R.Z. Valiev, chapter 5 – Yu.R. Kolobov, G.P. Grabovetskaya, K.V Ivanov, E.V. Naidenkin, chapter 6 – E.F. Dudarev, O.A. Kashin, Yu.R. Kolobov, and chapter 7 – R.Z. Valiev, A.P. Zhilyaev.

Some of the concepts and considerations proposed by the authors have already been discussed and published in earlier monographs [1,5,8]. These data have been supplemented by the latest results.

References

1. Kolobov Yu.R., Diffusion-controlled processes at grain boundaries and plasticity of metallic polycrystals, Novosibirsk, Nauka, 1998.
2. Fischer J.G., J. Appl. Phys., Vol. 22, No. 1, 74.
3. Mishin Yu.M., Razumovskii I.M., in: Structure and properties of internal interfaces in metals, Moscow, Nauka, 1988.
4. Kaur I., Mishin Yu., Gust W., Fundamentals of grain and interphase boundary diffusion, 3rd edition, Chichester, John Wiley and Sons Ltd, 1995.
5. Kaibyshev O.A., Valiev R.Z., Grain boundaries and properties of metals, Moscow, Metallurgiya, 1987.
6. Orlov A.N., Perevezentsev V.N., Rybin V.V., Grain boundaries in metals, Moscow, Metallurgiya, 1980.
7. Gleiter H., Progress Mater. Sci., 1989, Vol. 33, No.4, 223.
8. Valiev R.Z., Aleksandrov I.V., Nanostructured materials produced by severe plastic deformation, Moscow, Golos, 2000.
9. Investigation and applications of severe plastic deformation, Proc. NATO ARW, Moscow, August 2–8 (Eds. T.C. Lower, R.Z. Valiev), Dordrecht, Kluewer Publ., 2000.
10. Panin B.E., Likhachev V.A., Grinyaev Yu.B., Structural levels of deformation of solids, Novosibirsk, Nauka, 1985.

CHAPTER 1

METHODS OF PRODUCING BULK NANOSTRUCTURED METALS AND ALLOYS

Several methods have been developed for the production of nanostructured (NS) materials. They include the methods of powder metallurgy, in particular, the consolidation of ultrafine powders, produced by gas condensation in the atmosphere of an inert gas [1, 2] or by a plasma-chemical process [2], aerosol [3] and chemical synthesis [4], and also by the milling of powders in a ball mill [5, 6]. Some of these methods are used efficiently for the production of bulk nanostructured materials. It is mainly gas condensation with subsequent compacting and treatment of powders in a ball mill with subsequent consolidation. These methods represent the basis of a large number of investigations of the structure and properties of nanocrystalline and nanophase materials. At the same time, there are still problems in the development of these methods associated with the retention of some residual porosity in compacting, contamination of the specimens in the preparation or consolidation of powders, the increase of the geometrical dimensions of the produced specimens, and practical application of these methods.

Many of these problems can be overcome using the methods of treatment referred to as severe plastic deformation (SPD) [7, 8]. The task of SPD methods is the formation of nanostructures in thick metallic specimens and components by high deformation at a low temperature. For example, as a result of cold rolling or drawing [9–11], it is possible to refine greatly the structure of metals. However, the produced structures are usually cellular or contain dislocation sub-boundaries with small-angle disorientation. The SPD methods make it possible to produce ultrafine structures of the grain type with the grain size of the order of several fractions of a micron [8, 12]. Special methods of mechanical deformation have been developed for the practical application of these methods, including severe torsion and high-pressure, equal-channel angular pressing, etc. Using these methods, it has been possible to produce bulk nanostructured specimens and blanks from greatly differing metals and alloys, including many industrial alloys and intermetallics.

The main methods of producing bulk nanostructured materials will be examined and their advantages and shortcomings evaluated.

1.1. GAS CONDENSATION OF POWDERS AND CONSOLIDATION

The method of production of nanostructured materials, proposed by Gleiter et al [1, 13, 14] is widely known and very popular. This method uses the procedure of evaporation and condensation of atoms for the formation of nanocrystalline (NC) particles deposited on the cold surface of a spinning cylinder; evaporation and condensation are carried out in the atmosphere of a rarefied inert gas, usually helium. The variation of the type of working atmosphere at the same gas pressure (transition from helium to xenon, i.e. from a gas with low-density to a high-density gas) is accompanied by an increase of the particle size by several factors. Usually, the particles of the surface condensate are faceted. In the same conditions of evaporation and condensation, the metals with a higher melting point form smaller particles. The deposited condensate is removed with a special scraper from the surface of the cylinder and collected in a collector. After pumping away the inert gas, preliminary (at a pressure of approximately 1 GPa) and final (pressure up to 10 GPa) pressing of nanocrystalline powder is carried out in vacuum (Fig. 1.1). This results in the formation of sheets with a diameter of 5–15 mm, with a thickness of 0.2–3.0 mm and density of 70–90% of the theoretical density of the appropriate material (up to 97% for nanocrystalline metals and up to 85% for nanoceramics [15]). The compacted nanomaterials, produced by this method, consist of the particles with the mean size of 1–2 up to 80–100 nm, depending on the evaporation and condensation conditions.

Elimination of contact with the surrounding medium in producing the nanopowder and its pressing make it possible to avoid contamination of compacted nanocrystalline specimens. This is very important in examining the nanostructure of metals and alloys. The

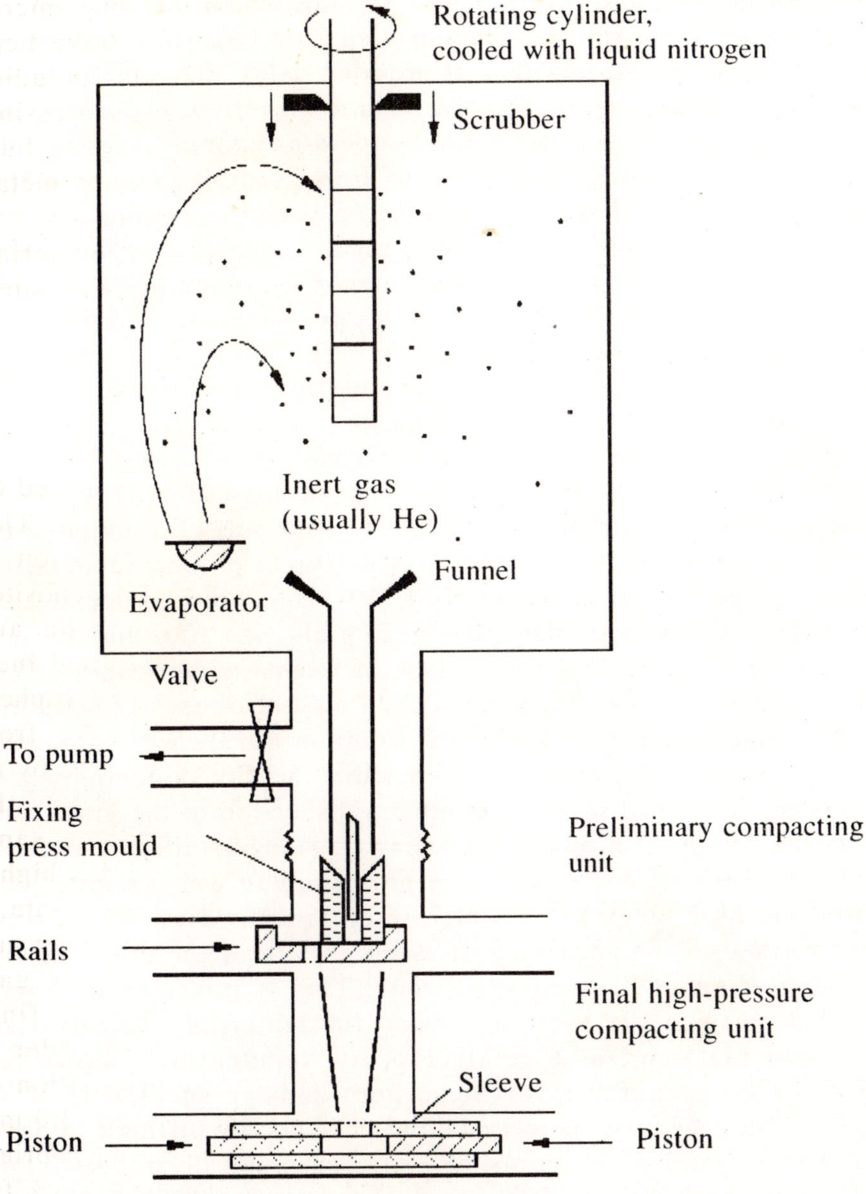

Fig. 1. Diagram of equipment for producing bulk nanocrystalline materials by evaporation, condensation and compacting (adapted from [1]).

equipment described in [1, 13, 14] may be used for producing compact nanocrystalline oxides and nitrides: in this case, the metal is evaporated in the oxygen- or nitrogen-containing atmosphere. The porosity of ceramics, produced by compacting the powders, is associated with triple joints of the crystallites. A decrease of the dispersion of the powder is accompanied by a large decrease of the compactibility of the powders in pressing using the same pressure [16]. The decrease and a more uniform distribution of porosity are achieved by pressing at a higher temperature which does not yet lead to intensive recrystallisation. For example, conventional sintering of a high-dispersion powder of zirconium oxide with the particle size of 40–60 nm at 1370 K results in a relative density of 72% at the mean grain size of the sintered specimen of 30 nm; hot pressing at the same temperature and a pressure 1.6 GPa produces a sintered material with a relative density of 87% and a mean grain size of 130 nm [17]. A decrease of sintering temperature to 1320 K and an increase of sintering time to 5 h produced compacted zirconium oxide ZrO_2 with a relative density greater than 99% and a mean grain size of 85 nm [18]. The authors of [19] used hot pressing of a titanium nitride powder (d ~80 nm) at a temperature of 1743 K and a pressing pressure of 4 GPa to produce compacted specimens with the density equal to 90% of theoretical density but (according to diffraction data) the mean grain size after hot pressing was not smaller than 0.3 μm because of intensive recrystallisation. The experimental results show [20] that the densest specimens (with a relative density of 98%) of titanium nitride were produced by sintering of specimens pressed from the finest nanopowders (diameter ~8–25 nm) with the minimum dispersion of the grain size.

On the whole, to produce high-density nanocrystalline materials, especially ceramic materials, it is promising to use pressing with subsequent high-temperature sintering of nanopowders. When using these methods, it is necessary to avoid growth of the grains in the sintering stage of pressed specimens. This is possible at a high initial density of the pressings when the sintering processes are relatively fast, and at a relatively low temperature (less than 0.5 T_m). The preparation of these high-density specimens is an important task because nanocrystalline powders are difficult to press and the commercial methods of static pressing do not result in sufficiently high density. The physical reason for the low pressing capacity of the nanopowders are the interparticle adhesion forces, whose relative value rapidly increases with a decrease of the particle size.

In compacting nanocrystalline powders, good results have been obtained by the magnetic-pulse method proposed by the authors of [21, 22]. The method of magnetic pulsed pressing makes it possible to generate pulsed compression waves with the amplitude up to

5 GPa and the duration of several microseconds. This method is based on the concentration of the force effect of the magnetic fields of powerful pulsed currents; it makes it possible to control quite simply the parameters of the compression wave, is ecologically clean and considerably safer than the dynamic methods using explosive substances.

1.2. BALL MILLING AND COMPACTING OF POWDERS

The method of ball milling is based on mechanical treatment of solid mixtures resulting in refining and plastic deformation of substances, acceleration of mass transfer, and also mixing of the components of the mixture on the atomic level, with the activation of the chemical interaction of solid reagents [23–25]. A stress field forms in the near-contact regions of the solid substance as a result of the mechanical effect. The relaxation of the stress field may take place by the generation of heat, the formation of a new surface, appearance of various defects in the crystals, and excitation of chemical reactions in the solid phase. Schematically, this method can be represented as shown in Fig. 1.2 [26]. The preferential direction of relaxation depends on the properties of the substance,

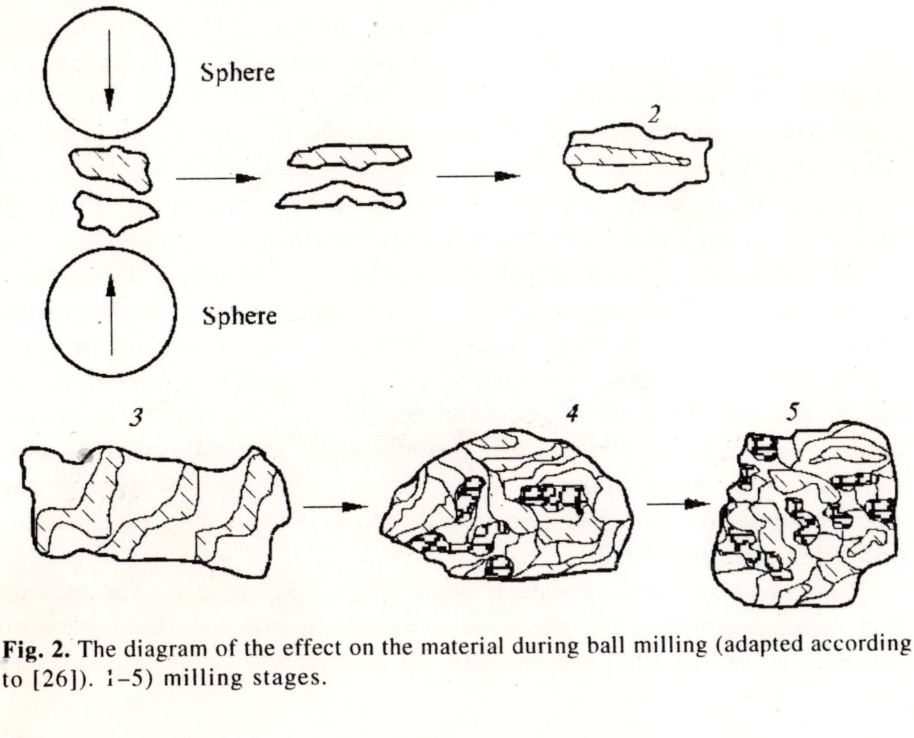

Fig. 2. The diagram of the effect on the material during ball milling (adapted according to [26]). 1–5) milling stages.

the loading conditions (the power of supplied energy, the ratio between pressure and shear), the size and form of the particles. The increase of the power of the mechanical pulse and the duration of the effect result in a gradual transition from relaxation by the generation of heat to the relaxation associated with the failure, dispersion and plastic deformation of the material and the appearance of amorphous structures of different nature. Finally, a channel of relaxation of the stress field may be a chemical reaction initiated by various mechanisms, such as the direct excitation and fracture of bonds, taking place in the crack tip, local thermal heating, emission-free breakdown of excitons, etc.

The mechanical effect in the milling of the materials is pulsed and, consequently, the stress field and its subsequent relaxation do not take place over the entire period of holding the material in the reactor but only at the moment of collision of the particles and a short time after the collision. For this reason, in mechanochemical synthesis it is important to take into account the nature of formation of the stress field and the kinetics of subsequent relaxation processes. The mechanical effect is not only pulsed but also localised because it does not take place in the entire mass of the solid substance but only in the areas of formation and subsequent relaxation of the stress field.

Mechanical milling is the most productive method of producing large quantities of nanocrystalline powders from different materials: metals, alloys, intermetallics, ceramics, composites. Mechanical attrition and mechanical fusion can be used to ensure complete solubility in the solid state of elements whose mutual solubility in the equilibrium conditions is extremely low [27, 28].

In the mechanical milling of powders, deformation is initially localised in slip bands containing a large number of dislocations with a high density. When reaching a specific stress level, these dislocations annihilate and recombine with low-angle boundaries separating the individual grains; the abrasion stage is already characterised by the formation of grains with a diameter of 20–30 nm and the number of these grains increases during attrition. In the next stage of attrition, the orientation of the individual crystals in relation to each other becomes random because of slip at the grain boundaries. This behaviour in attrition is typical of BCC metals and intermetallics [28].

Planetary, ball and vibration mills (Fig. 1.3) are used for milling and mechanochemical synthesis. The mean size of the produced powder is in the range from 200 to 50–10 nm. For example, in milling in a ball mill of α-FeB boride it was possible to produce a powder of α-FeB with the mean size of crystallites of approximately 8 nm [29]. The mechanical treatment of barium titanate $BaTiO_3$ in a planetary mill resulted in the formation of nanocrystalline powder with a mean particle size of 5–25 nm [24].

During milling, the material is subjected to severe high-rate deformation. The higher level of internal stresses forms as a result of the high-density of the dislocations, disclinations, vacancies and other crystal lattice defects, introduced during deformation [30]. The dislocation clusters undego recovery and form a cellular ordered structure and deformation slip bands which may transform with an increase of the degree of cumulated strain into new high-angle grain boundaries. Regardless of the recovery of the structure taking place at the reduced temperatures, the level of the internal stresses remains high. Figure 1.4 shows the variation of the mean grain size (*a*) and internal stresses (*b*) as a function of milling time, and also the dependence of the final value of the mean grain size (Fig. 1.4*c*) as a function of melting point for several FCC and

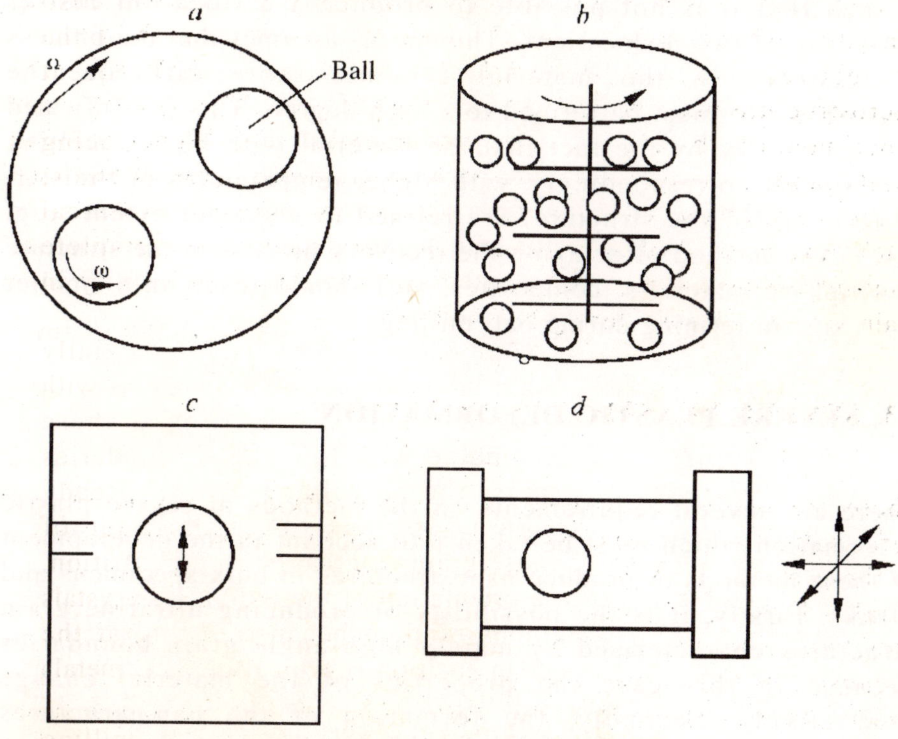

Fig. 3. The types of ball milling machines. *a*) planetary; *b*) attritor; *c*) one-dimensional vibration machine; *d*) three-dimensional vibration machine.

BCC metals. After ~10 hours of milling, the mean grain size becomes optimum and the further effect does not result in any refining of the structure. This time is also critical for the internal stress which starts to decrease after 10 hours. A decrease of the internal stress during continuation of the milling process is explained by the fact that a decrease of the mean grain size in the nanometric region (<10 nm) requires a very high level of external stresses for the generation of new dislocations, and dislocation clusters may consist of 2–3 dislocations. Consequently, the minimum possible grain size can be determined from the following relationship [26]:

$$d_{cr} = \frac{3 \cdot G \cdot b}{(1 - \nu_p) \cdot H}, \tag{1.1}$$

Here G is the shear modulus, b is the Burgers vector, ν is the Poisson coefficient, H is the hardness of the material. This formula determines the level of stresses built up during deformation which is such that it is not possible to produce a dislocation cluster consisting of two dislocations. This model assumes that the process of recovery in the material becomes more difficult, the microstructure must be refined to a high degree. This is confirmed experimentally by the fact that the material with higher melting points (and, correspondingly, with higher temperatures of the start of recovery of the structure) are refined to a greater extent (Fig. 1.4*c*). Any method of delaying the recovery process in the structure (multiphase materials, composites, etc) should result in a smaller grain size in refining during ball milling.

1.3. SEVERE PLASTIC DEFORMATION

There are several requirements on the methods of severe plastic deformation which must be taken into account in the development of these methods to produce nanostructures in bulk specimens and blanks. Firstly, it is the possibility of producing ultrafine grain structures characterised by mainly high-angle grain boundaries because in this case the properties of the material change qualitatively. Secondly, the formation of the nanostructures homogeneous over the entire volume of the specimens is another requirement. This is essential for ensuring the stability of the properties of the produce materials. Thirdly, the specimens should

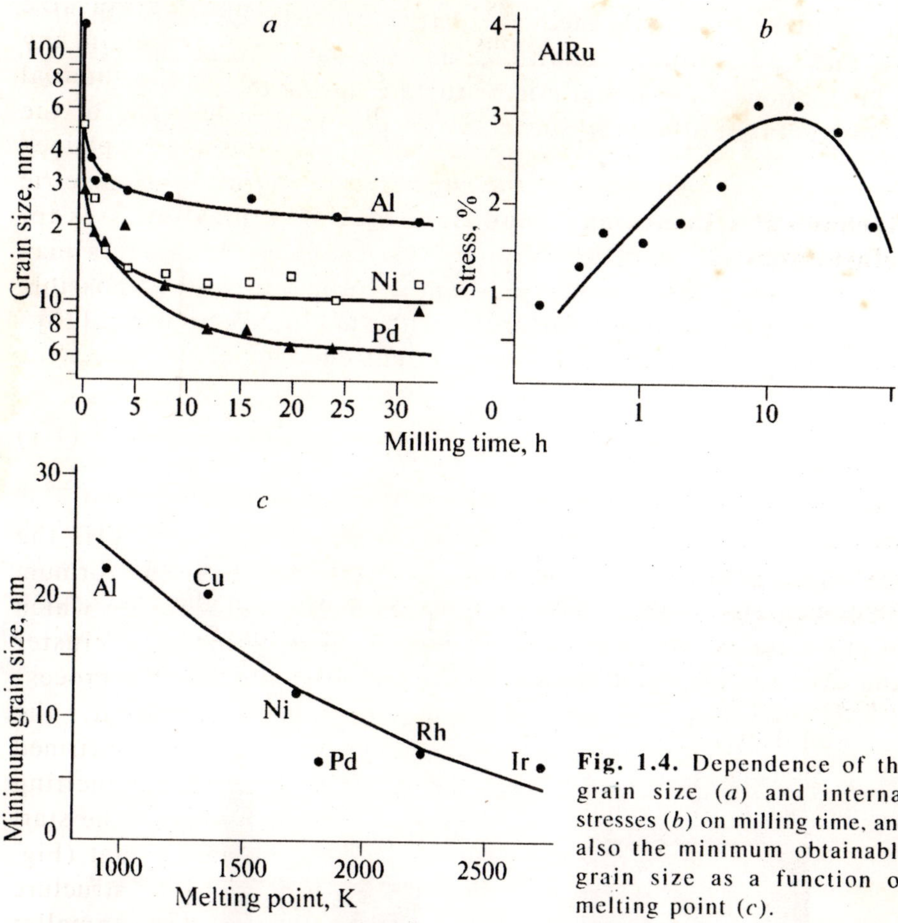

Fig. 1.4. Dependence of the grain size (a) and internal stresses (b) on milling time, and also the minimum obtainable grain size as a function of melting point (c).

not contain any mechanical damage or failure, regardless of severe deformation. This requirements can be realised by the application of conventional methods of pressure treatment of metals, such as rolling, drawing or extrusion. The nanostructures in the bulk specimens can be produced using special mechanical deformation methods resulting in high strains in the materials at a relatively low temperature, and it is also essential to determine the optimum treatment conditions of the materials. At present, the majority of the results have been obtained using two methods of severe plastic deformation – high pressure torsion, and equal-channel angular pressing. Studies have also been published on the production of nano- and sub-microcrystalline structures in a number of metals and alloys using uniform forging [31, 32] and equal-channel angular drawing [33].

In this section, attention will be given to the problems of

application of the main methods of severe plastic deformation, their schemes and optimum conditions, and the data will be presented on the evolution of the initial microstructure during the formation of the nanostructured all condition in severe plastic deformation.

Mechanical schemes and conditions of severe plastic deformation

The main methods of achieving high deformation with true degrees of 10 or more without failure of the specimens include torsion and high-pressure (Fig. 1.5) and equal-channel angular pressing (Fig. 1.6).

Deformation by high-pressure torsion

The systems in which torsion deformation is carried out under high-pressure were used for the first time in [34, 35]. The design of the systems is the development of the well-known concept of the Bridgman anvil [36]. These systems were used for the examination of phase transformations in the conditions of severe deformation [34] and also for the examination of the evolution of the structure and

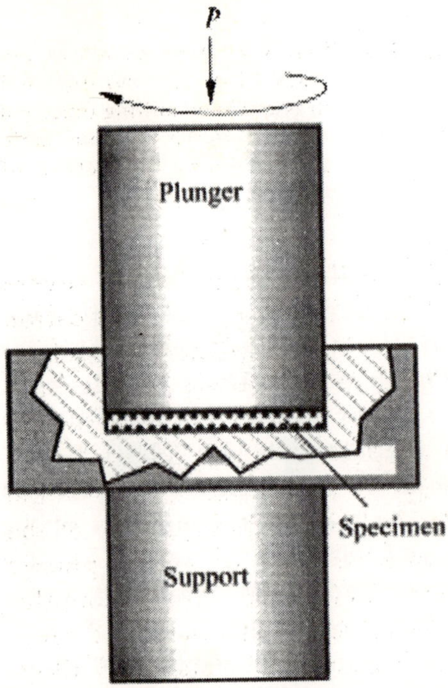

Fig. 1.5. Severe plastic deformation by torsion at high pressure.

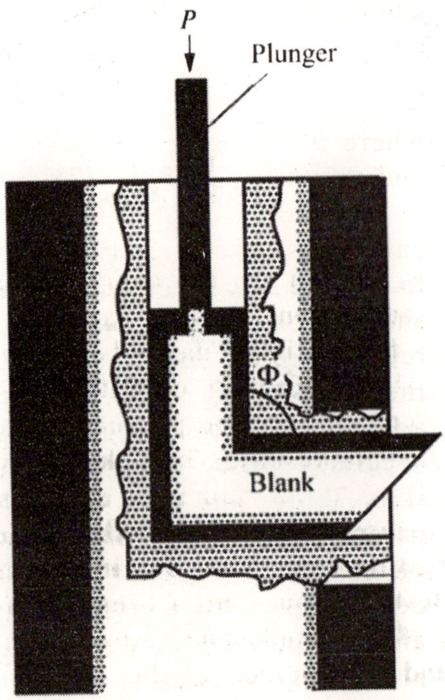

Fig. 1.6. Severe plastic deformation by equal-channel angular pressing.

changes of the recrystallisation temperature after high deformation [35]. A new and a very important moment has been the confirmation of the formation of nanostructures with non-equilibrium high-angle grain boundaries using severe deformation by torsion [7, 8, 12]. Consequently, the method was regarded as a new method of producing nanostructure materials. The mechanical aspects of severe deformation by torsion will be described.

In deformation by high-pressure torsion (Fig. 1.5) the resultant specimens are disc-shaped. The specimen is placed between strikers and compressed under applied pressure (P) of several gigapascals. The lower striker rotates and the surface friction forces result in the shear deformation of the specimen. The geometrical form of the specimens is such that the main volume of the material is deformed in the conditions of hydrostatic compression under the effect of applied pressure and the pressure from the side of the external layers of the specimen. Consequently, regardless of the high degree of deformation, the specimen does not fracture.

Various relationships are used for calculating the degree of deformation in high-pressure torsion. For example, in [37] of the true logarithmic strain ε was calculated using the following equation:

$$\varepsilon = \ln(\vartheta \cdot r / h), \tag{1.2}$$

where ϑ is the angle of rotation in radians; r and h are the radius and thickness of the disc, respectively.

This equation is identical with the relationship used for the calculating the true strain of the specimens, subjected to tensile loading. However, if in tensile loading this equation is physically substantiated, in torsion it is not. In particular, according to this relationship, in torsion under pressure, the logarithmic strain around the perimeter of the typical specimens with a diameter of 20 mm and a thickness of 1 mm is 6 and along the perimeter of the specimens with a diameter of 10 mm and a thickness of 0.2 mm it is 7. At the same time, in the centre of the specimens this parameter is equal to 0. According to the results of a large number of investigations, in the application of this method of severe plastic deformation, the structure in the central part of the specimens is also refined after several revolutions and is usually homogeneous along the radius of the specimens. This is confirmed by similar values of microhardness at different points (both in the centre and at the periphery of deformed specimens). The strain in this method was also calculated using the equation:

$$\varepsilon_c = 2\pi \cdot R \cdot \frac{N}{h}, \tag{1.3}$$

used in conventional torsional deformation for calculating the shear strain at the distance R from the axis of the disc-shaped specimen. Here N is the number of revolutions, h is the thickness of the specimen. To compare the torsional shear strain with the strain in other deformation methods, the former is usually transformed to the so-called equivalent strain ε_{eq}. According to the von Mises criterion

$$\varepsilon_{eq} = \frac{\varepsilon_c}{\sqrt{3}}. \tag{1.4}$$

Two comments should be made regarding equation (1.3) [38]: 1) calculations carried out using this equation lead to a conclusion that the strain should change in a linear manner from zero in the centre of the specimen to the maximum value at the ends of the diameter of the specimen but, as already mentioned, this is often not detected in the experiments; 2) in deformation, the initial thickness of the specimen is approximately halved under the effect of high compressive pressure and, consequently, the use of the initial thickness of the specimen as the value h reduces the calculated values of the strain in comparison with the true values.

Both these comments indicate that the strains, calculated using the previously mentioned equations, are only approximately equal to the actual strain. In addition to this, in the process of severe plastic deformation, the nanostructures form not only under the effect of external but also internal stresses. At the same time, there is no distinctive relationship between the internal stresses and true strains. This confirms the formation of a usually homogeneous structure along the diameter of the specimens, subjected to severe plastic deformation by torsion, although in accordance with equations (1.2) and (1.3) no significant refining of the microstructure should take place in the centre of the specimens. Therefore, in examination of the process of evolution of the microstructure in the course of severe plastic deformation by torsion it is more accurate to examine the number of revolutions, and not the value of the strain calculated using analytical expressions. This position becomes especially important in the treatment of low-deformability or brittle materials where sliding between the strikers and the specimen or cracking of the latter is not possible. To prevent this from taking place, it is necessary to increase the applied pressure which creates additional technical difficulties in the selection of a stronger material of the strikers and in the optimisation of the design of equipment.

The specimens, produced by severe plastic deformation in torsion, have the form of discs with a diameter of 10–20 mm, thickness 0.2–0.5 mm. The structure is greatly refined already after deformation by half revolution [38] but the formation of the uniform nanostructure usually requires deformation in several revolutions.

Recent investigations have also shown that severe plastic deformation in torsion may be used efficiently not only for refining the structure but also as a method of consolidation of the powders [39–45]. It has been found that high pressures, equal to several GPa, in deformation by torsion at room temperature may result in a relatively high density (close to 100%) of the produced nanostructured disc specimens. To produce the specimens by consolidation by severe plastic deformation in torsion, it is recommended to use not only conventional powders but also powders processed in a ball mill.

To illustrate this, let us discuss the example of compacting a nanostructured nickel powder, produced in a ball mill, by compacting in severe plastic deformation by torsion [40]. The experimental results show that the density of the produced specimens is close to 95% of the theoretical density of thick coarse-crystalline nickel. In this case, the specimens did not contain porosity visible in a transmission electron microscope and the mean grain size was very small (approximately 17 nm). Consequently, the volume occupied by the grain boundaries was relatively large. The authors of [40] assumed that the theoretical density of the specimens decreases owing to the fact that the grain boundaries in the materials with a very small grain size and extensive distortion of the crystal lattice are characterised by reduced atomic density. It is also interesting to note that the microhardness of the nickel specimens, produced by consolidation by severe plastic deformation, was 8.60±0.17 GPa. This is the highest microhardness of nanostructured nickel, mentioned in the literature.

Deformation by equal-channel angular pressing

The method of equal-channel angular pressing, realising the deformation of thick specimens by simple shearing, was developed in the 70s by Segal *et al* in order to subject materials to plastic deformation without changing the cross-section of the specimens. This enables secondary deformation of the materials [41, 42] (Fig. 1.6). At the beginning of the 90s, Valiev *et al* developed this method and used it for the first time as the method of severe plastic deformation to produce structures with sub-microcrystalline and nanometric grain sizes [8, 46–48]. In these experiments, initial blanks with the circular or square cross-section were cut from bars 70–100 mm long. The diameter of the cross-section or its diagonal usually do not exceed 20 mm. To apply equal-channel angular pressing, the component is extruded several times in special equipment through two channels with the same cross-section intersecting each other usually under the angle of 90°. In low-deformability materials, deformation takes place at elevated temperatures.

When the external angle $\psi = 0°$, and the internal angle Φ is arbitrary (Fig. 1.7), according to [49, 50] the increase of the strain $\Delta\varepsilon_i$ in shearing during each pass through the channels may be calculated from the following equation:

$$\frac{P}{\sigma_T} = \Delta\varepsilon_i = \frac{2}{\sqrt{3}} \cdot \mathrm{ctg}(\Phi / 2), \tag{1.5}$$

where P is applied pressure; σ_T is the flow stress of the deformed material.

Since in equal-channel angular pressing the component is extruded several times through the intersecting channels, the total strain is:

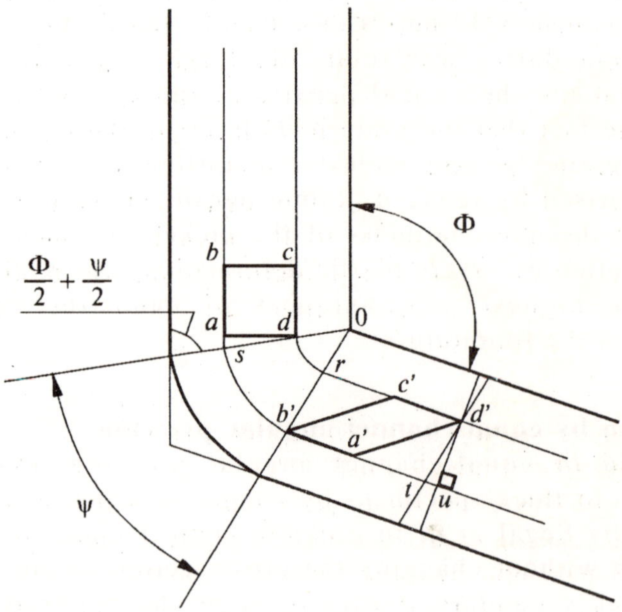

Fig. 1.7. Principles of equal-channel angular pressing.

$$\varepsilon_N = N \cdot \Delta\varepsilon_i, \tag{1.6}$$

where N is the number of passes.

The more general relationship, which makes it possible to calculate the strain in the specimen in equal-channel angular pressing in N passes, has the following form [15]:

$$\varepsilon_N = N\left[\frac{2\cdot \mathrm{ctg}(\Phi/2+\psi/2)+\psi\cdot \mathrm{cosec}(\Phi/2+\psi/2)}{\sqrt{3}}\right]. \tag{1.7}$$

This relationship shows that at the most frequently used angles $\Phi = 90°$ and $\psi = 20°$, each pass corresponds to the additional strain equal to approximately 1.

In equal-channel angular pressing, the formation of the structure is controlled by the direction and number of passes of the blank through the channels. The authors of [49, 52–54] investigated various routes of the blanks (Fig. 1.8): the orientation of the blank is constant in each pass (route *A*); after each pass, the blank rotates through the angle of 90° around its longitudinal axis (route *B*); after each pass, the blank rotates through the angle of 180° along its longitudinal axis (route *C*). These routes differ in the direction of shearing in repeated passage of the blank through the intersecting channels and lead to changes in the shape of the spherical cell in the body of the blank during equal-channel angular pressing.

During the first pass, as a result of simple shearing in equal-channel angular pressing, the cell in the area of interaction of the channels becomes ellipsoidal (Fig. 1.9*a*). Subsequent passes in the case of route *A* increase the length of the axis 1, and the ellipsoids is elongated. In this case, the direction of shearing rotates through the angle 2Φ around the axis perpendicular to the longitudinal section of the channels. This is clearly demonstrated in Fig. 1.7. The secondary passage in the case of route *B* changes the direction of shearing; in this case, the shearing plane rotates through the angle of 120° (at 2Φ = 90°) (Fig. 1.9*c*) and the secondary passage results in shearing in the same plane but in the opposite direction *C* (Fig. 1.9*c*). In this case, the cell again becomes spherical.

The application of all three routes results in a rapid increase of the yield and tensile strength values of the processed material which reach saturation already after several passes [50]. It was shown in [46] that the first three deformation cycles of the specimens of

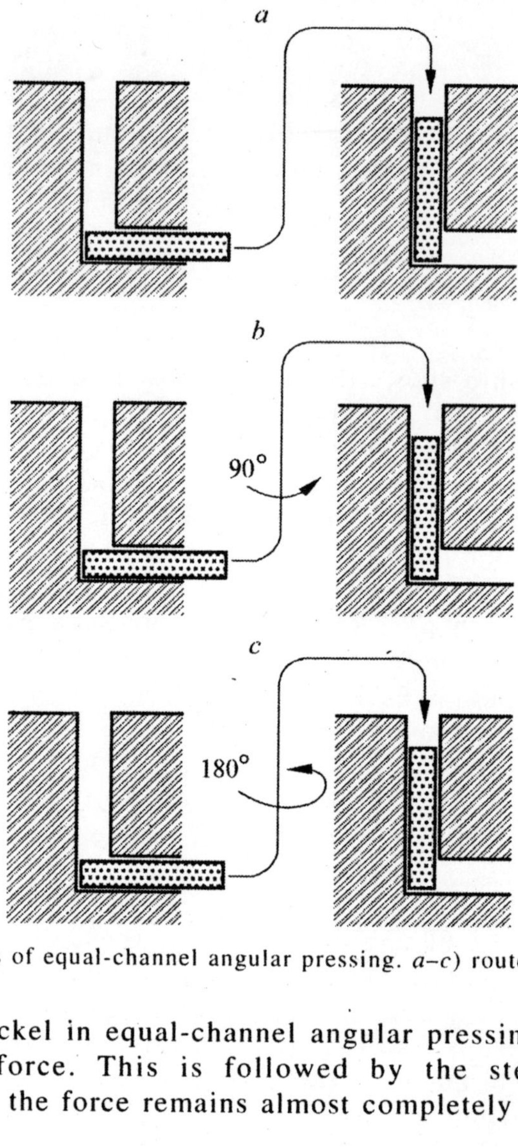

Fig. 1.8. Variants of equal-channel angular pressing. *a–c*) routes *A*, *B*, *C* [51].

copper and nickel in equal-channel angular pressing increase the deformation force. This is followed by the steady stage of hardening and the force remains almost completely constant.

Uniform forging

Another method of producing nanostructures in the thick specimens is uniform forging, proposed by Salishchev et al [31, 55, 60–63]. The process of uniform forging is usually accompanied by dynamic recrystallisation. The method of uniform forging is based on the application of multiple repetition of the free forging operations: upsetting–drawing with the change of the axis of applied deformation force. The homogeneity of deformation in this technological method is lower in comparison with equal-channel

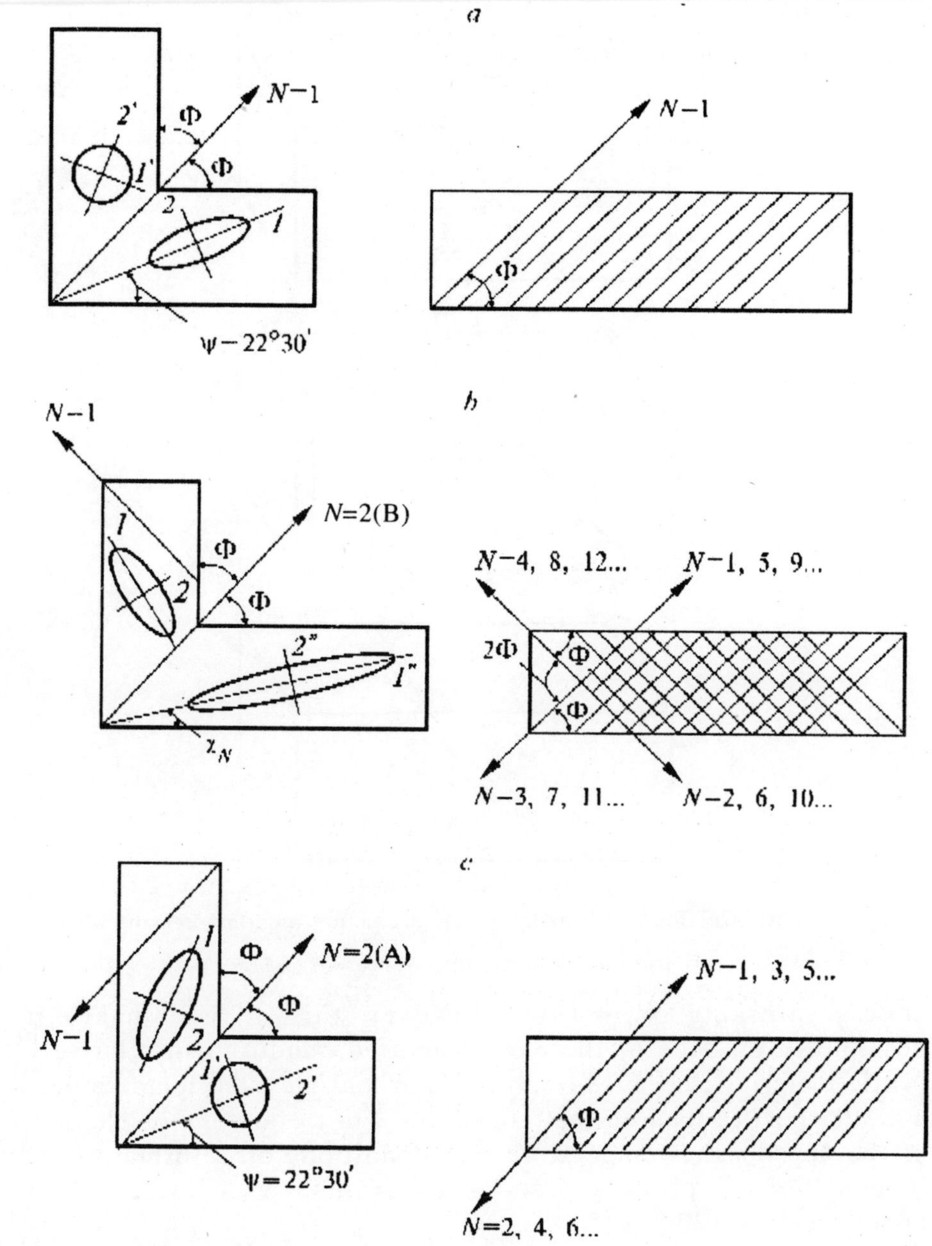

Fig. 1.9. The conditions of simple shearing in equal-channel angular pressing [42]. *a*) single-cycle deformation; *b*) multi cycle deformation, route *A*, *c*) the route *C*.

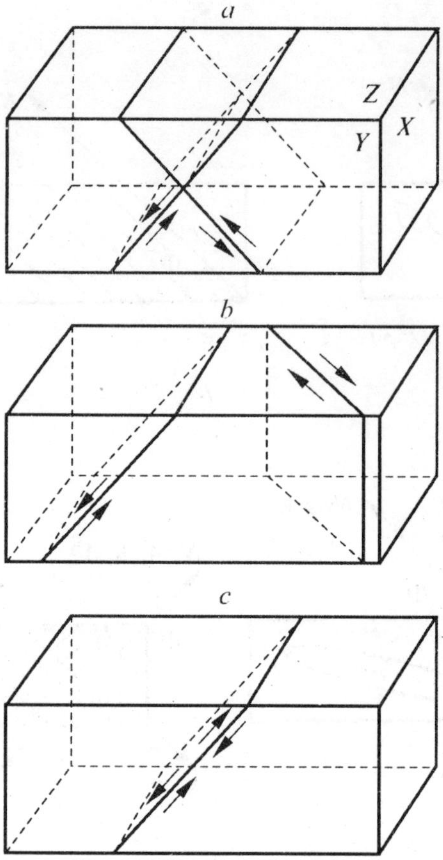

Fig. 1.10. The direction of shearing in equal-channel angular pressing with the routes *A (a)*, *B (b)* and *C (c)* [51].

angular pressing or torsion. However, this method makes it possible to produce of the nanostructured condition in relatively brittle materials, because treatment starts at elevated temperatures and the specific load on the tool is low. For example, the selection of the appropriate temperature-rate conditions of deformation. If it possible to produce very fine grains, with the size of approximately 100 nm [55–63].

CHAPTER 2

RELATIONSHIPS AND MODELS OF THE PROCESS OF GRAIN BOUNDARY DIFFUSION IN POLY- AND BICRYSTALS

The special attention given to the investigations of the diffusion properties of nanocrystalline (NC) and nanostructured (NS) metals and alloys is continuing. This is due to the fact that the diffusion properties play an important and sometimes controlling role in obtaining the unique properties of plasticity and superplasticity, processes of degradation and recovery of the structure of nanomaterials. The examination of diffusion in these materials has become intriguing and very important because of a number of experimental studies in which it was found that the diffusion coefficients (mainly grain boundary diffusion) in NC and NS metals and alloys are several orders of magnitude higher than the coefficients in coarse-grained materials [64, 167].

However, to obtain reliable results in the determination of diffusion coefficients at the grain boundaries (GB), it is often necessary to overcome large problems in the selection of an adequate model of grain boundary diffusion for analysis of the experimental data. The small grain size, the possibility of overlapping of diffusion flows of atoms from the adjacent boundaries, relaxation and migration of GB during diffusion annealing and, above all, the heterogeneity of the structure of the volume of the grains strongly affect the kinetics of diffusion processes. Consequently, the results of analysis of the experimental data may greatly differ.

In this chapter, we discuss the latest achievements in the development of analytical models of grain boundary diffusion.

Special attention is given to the relationships and diffusion models in order to analyse this process in nanocrystalline and nano-structured materials starting with the well-known Fisher solution for diffusion along a single grain boundary and Harrison's classification for grain boundary diffusion in a polycrystal to current views regarding diffusion in metals with the heterogeneous grain structure.

2.1. HARRISON CLASSIFICATION OF GRAIN BOUNDARY DIFFUSION REGIMES

For consecutive theoretical examination of grain boundary diffusion it is important to mention the existence of stages of the process and the possibility of separating the main kinetic regimes. The realisation of a specific regime depends on both the parameters of the system itself and on the experimental conditions.

A classification of grain boundary diffusion was proposed for the first time in 1961 by Harrison who specified three kinetic regimes of this process: **A**, **B** and **C** [65–67].

According to Harrison, in kinetic regime **C**, the diffusion flow in a polycrystal propagates only along the grain boundary and diffusion in the volume is insignificant ('frozen'). The criterion for realisation of regime **C** is expressed by the condition:

$$(D_v t)^{1/2} < \delta / 20 \tag{2.1}$$

where δ is the diffusion width of the GB, D_v is the volume diffusion coefficient; t is time.

Kinetic regime **B** includes the diffusion taking place both at the boundary and in the volume of the grain, but diffusion flows from the adjacent GBs do not overlap and, consequently, GB may be regarded as isolated. This is also found when the following condition is fulfilled:

$$(D_v t)^{1/2} < d/20, \tag{2.2}$$

where d is the grain size.

Kinetic regime **A** belongs to the limiting cases of long annealing times, small grain sizes or relatively high volume diffusion coefficients (when the value of the volume diffusion is comparable or slightly higher than that of the grain boundary diffusion coefficient, as in the case of diffusion of interstitial impurities in some metals). The path of volume diffusion is comparable with the path of grain boundary diffusion and the main saturation of the volume with the diffusant is determined by diffusion directly from the surface. It may appear that the diffusion path in the volume is considerably longer than the distance between the grain boundaries and the fields of volume diffusion from adjacent boundaries overlap each other. Each atom of the diffusant moves along several GBs and in the volume between them. Consequently, atomic transfer can be described by the effective diffusion coefficient (according to Hart) [66–68]:

$$D_{eff} = \tau D_b + (1-\tau)D_v, \tag{2.3}$$

where D_b is the grain boundary diffusion coefficient; τ is the fraction of atomic sites for GB (or this is equivalent to the volume fraction of the matter of the grain boundaries in a polycrystal).

The criterion for the realisation of regime **A** was introduced by Harrison in analysis of the problem of grain boundary diffusion in a specimen consisting of spherical grains with diameter d, and the grain boundary diffusion coefficient was regarded as considerably higher than the volume diffusion coefficient. Harrison showed that the polycrystal can be regarded as a homogeneous medium with the effective diffusion coefficient D_{eff} (equation (2.3)), if the following condition is fulfilled:

$$(D_v t)^{1/2} \geq 150d. \tag{2.4}$$

Although in microscopic examination of regime **A** by Hart (2.3) the experimental data are described quite efficiently using the unique effective diffusion coefficient, a complete and vertical model for regime **A** has not been constructed.

For more detailed examination of the grain boundary diffusion processes (and, in particular, the conditions **B** and **C**, which are of greatest practical interest), it is necessary to develop physical models of the GB and a phenomenological description of grain boundary diffusion.

2.2. FISHER MODEL

The Fisher model is the model that is generally recognised and used most widely for processing the experimental data for grain boundary diffusion [66, 67, 69]. Fisher examined diffusion in a bicrystal with a flat GB normal to the surface. The boundary has the form of a homogeneous isotropic sheet with thickness δ (Fig.2.1). The grain boundary diffusion coefficient D_b is assumed

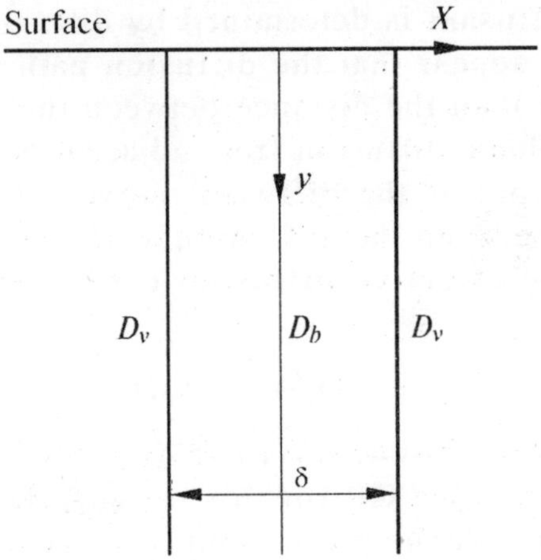

Fig.2.1. The Fisher model of an isolated grain boundary [69]. D_v and D_b are the coefficients of volume and grain boundary diffusion; δ is the diffusion width of the GB.

to be considerably higher than the volume diffusion coefficient D_v.

The kinetic equations for the examined model are presented in the following form:

$$\frac{\partial C_v}{\partial t} = D_v \left(\frac{\partial^2 C_v}{\partial x^2} + \frac{\partial^2 C_v}{\partial y^2} \right), \tag{2.5}$$

$$\frac{\partial C_b}{\partial t} = D_b \frac{\partial^2 C_b}{\partial y^2} + \frac{2D_v}{\delta} \frac{\partial C_v}{\partial x} \bigg|_{x=\delta/2}. \tag{2.6}$$

They describe the concentration of an impurity in the body of the grain C_v and at the boundary C_b under the boundary condition

$$C_b \big|_{x=\delta/2} = C_v \big|_{x=\delta/2}. \tag{2.7}$$

In the model it is assumed that the coefficients of volume and grain boundary diffusion D_v and D_b are independent of the coordinate and time. In equation (2.6) the first term on the right-hand side describes diffusion along the GB, the second one the outflow of the impurity from the GB to the volume. It should be added that equation (2.6) may be derived from the diffusion equation

$$\frac{\partial \overline{C}_b}{\partial t} = D_b \nabla^2 \overline{C} \tag{2.8}$$

for the mean concentration $\bar{C}_b$ in the GB taking into account the equality of the flows:

$$D_v \frac{\partial C_v}{\partial x}\bigg|_{x=\delta/2} = D_b \frac{\partial C_b}{\partial x}\bigg|_{x=\delta/2} \tag{2.9}$$

Thus, averaging (2.8) in respect of x in the ranges $(-\delta/2, \delta/2)$, we have [67]

$$\frac{\partial \bar{C}_b}{\partial t} = \frac{D_b}{\delta} \int_{-\delta/2}^{\delta/2} \frac{\partial^2 C_b}{\partial x^2} dx + D_b \frac{\partial^2 \bar{C}_b}{\partial y^2} =$$

$$= \frac{D_b}{\delta} \cdot \frac{\partial C_b}{\partial x}\bigg|_{x=\delta/2} - \frac{D_b}{\delta} \frac{\partial C_b}{\partial x}\bigg|_{x=-\delta/2} + D_b \frac{\partial^2 \bar{C}_b}{\partial y^2}.$$

Using (2.9) we obtain

$$\frac{\partial \bar{C}_b}{\partial t} = D_b \frac{\partial^2 \bar{C}_b}{\partial y^2} + \frac{2D_v}{\delta} \frac{\partial C_v}{\partial x}\bigg|_{x=\delta/2}$$

It should be mentioned that the examined model of grain boundary diffusion permits the determination of the width of the GB on the basis of experimental data only for self-diffusion in pure materials. In hetero- and self-diffusion in multicomponent systems it is important to take into account the effect of segregation of impurities of the grain boundaries on the effective value of σ. The first attempt to take into account the segregation effects in mathematical analysis of grain boundary diffusion was made by V.S. Bokshtin, N.A. Magidson and E.L. Svetlov [67, 70] in 1958. The model shows that to take into account the segregation of the impurity at the grain boundaries, the value of δ in mathematical analysis should be replaced by $s\delta$ and, consequently, product D_b by product $s\delta D_b$, where s is the segregation coefficient [67]. A similar assumption was later proposed and justified by Gibbs [71] who took into account the segregation effects in the classic model of a homogeneous layer for grain boundary diffusion. In this case, the boundary diffusion (2.7) has the general form [67]:

$$C_b\big|_{x=\delta/2} = C_v\big|_{x=\delta/2}. \tag{2.7a}$$

The solution of the system of equations (2.5)–(2.6) depends on the selection of the source of the diffusant on the external surface.

Three models of the source can be considered at present: Whipple, Suzuoka and the model of fast surface diffusion (FSD) [65].

In the Whipple model [72] it is assumed that the concentration of the diffusant on the surface of the specimen C_0 remains constant during the diffusion experiment

$$C_0 = \text{const.} \tag{2.10}$$

The condition for realisation of this model is the large thickness of the coating h ($h \gg L$, where $L = 2\sqrt{D_v t}$ is the diffusion path in the volume) and fast (faster than grain boundary diffusion in the specimen) diffusion at the specimen–coating interface, i.e. the case found most frequently in metals $D'' \gg D_b \gg D_v$, where D'' is the diffusion coefficient.

In Suzuoka's model [73] it is assumed that at the initial moment of time, an infinitely thin layer of the diffusant with a surface density K is deposited on the surface of the specimen:

$$C(x, y, 0) = K\delta(y), \tag{2.11}$$

where δ (y) is the Dirac delta function.

During annealing the atoms of the diffusant are ideally reflected from the surface and this is expressed by the condition

$$\frac{\partial C(x,t)}{\partial y} = 0. \tag{2.12}$$

This model is valid if the diffusion permitivity of the surface D'' is close to that for volume ($D'' \sim D_v$) and the part of the boundary, adjacent to the surface and depleted in the diffusant because of high permeability, is not a sink for the diffusant.

In a case identical to the Whipple condition but with long annealing times, it is necessary to use the model of fast surface diffusion (FSD) model [74]. The following conditions were accepted in this model:

$$C(x, y, 0) = K\delta(y), \tag{2.13}$$

$$C(x, 0, t) = K / \sqrt{\pi D_v t}. \tag{2.14}$$

The physical meaning of selection of conditions (2.13) and (2.14) is discussed in [65, 67, 74].

The applicability of this or other models in relation to the

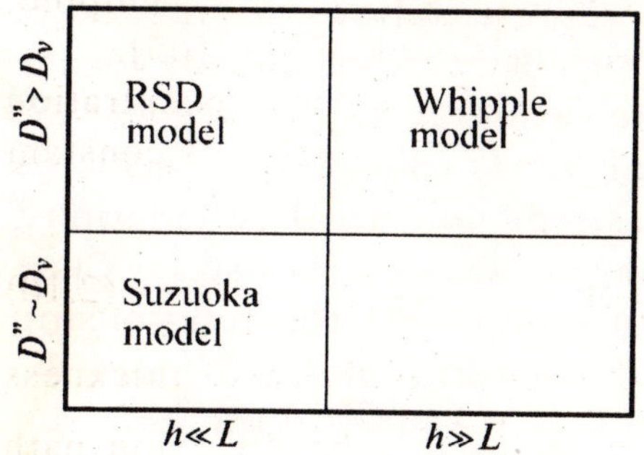

Fig.2.2. Conditions of applicability of different models of grain boundary diffusion in relation to the ratio of the coating thickness h and the diffusion path L, the coefficients of volume D_v, grain boundary D_b and surface D'' diffusion [74].

experiment conditions is shown in Fig.2.2.

2.3. EXACT SOLUTIONS OF DIFFUSION EQUATIONS IN THE FISHER MODEL

The system (2.5), (2.6) was solved for the first time by Fisher for a source of constant concentration [67, 69]. When deriving an analytical solution, Fisher proposed the absence of direct volume diffusion from the surface and quasi-equilibrium character of grain boundary diffusion expressed by the condition $\frac{\partial C_b}{\partial t}=0$ in equation (2.6). In this case, the boundary can be regarded as a source of constant concentration or diffusion into the volume. This gave the equation for the concentration of the impurity in the grain boundary and in the volume and also for layer concentration (the amount of the diffusant in the thin layer, parallel to the surface), determined using the equations [67]

$$\bar{C}(y,t) = C_b(y,t)\delta(y) + \int_{-\infty}^{-\delta/2} C_v dx + \int_{-\delta/2}^{\infty} C_v dx. \tag{2.15}$$

The main special feature of the solution is the linear dependence of ln C_b and ln $\bar{C}$ on the depth of penetration y, from which we can determine, using the angle of inclination $\frac{\partial \ln \bar{C}}{\partial y}$, only the product δD_b at the known value of D_v.

However, the exact solution of the system (2.5), (2.6) using the method of Fourier–Laplace transforms was obtained by Whipple for a source with constant concentration [72] and by Suzuoka for diffusion from an ‘instantaneous’ source [73]. Because of their cumbersome nature, the integral representations of the exact solutions proposed by Whipple and Suzuoka are not suitable for the analysis of the experimental data. However, taking into account the studies by Levin and MacAllum [75] who determined the dependence of ln on $y^{6/5}$ on the basis of numerical calculations, Suzuoka determined for an instantaneous source [76] and Le Clare for a source with constant concentration [77] the relationship between the angle of inclination of the curve ln $\bar{C}\,(y^{6/5})$ and the product δD_b [67]. For example, for a source with constant concentration

$$\delta D_b = 2(D_v / t)^{1/2} (\partial \ln \bar{C} / \partial y^{6/5})(0.78)^{5/3}. \tag{2.16}$$

Equation (2.16) is referred to as the Whipple–Le Clare equation.

The Suzuoka equations are characterised by a parametric dependence on the quantity $\beta = \dfrac{\delta D_b}{2D_v^{3/2} t^{1/2}}$ and were published in [67].

2.4. GENERAL CLASSIFICATION OF GRAIN BOUNDARY DIFFUSION REGIMES FOR A STATIONARY GRAIN BOUNDARY

Taking into account the exact solutions of the Fisher model by Whipple and Suzuoka, Razumovskii and Mischin [74, 78–81] proposed a general classification of the grain boundary diffusion regimes. They noted that in experiments only one of the diffusion regimes is often realised and they are separated on the basis of dimensionless parameters λ and ν. The exact solution can also be written in dimensionless parameters (λ and ν) and in the co-ordinates (u and ω) [74, 78–81]:

$$u = \begin{cases} 0 & -\delta/2 \le x \le \delta/2, \\ (x - \delta/2)/2\sqrt{D_v t}, & x > \delta/2, \\ -(x + \delta/2)/2\sqrt{D_v t}, & x < -\delta/2; \end{cases}$$

$$\omega = y / L_b;\ \mathrm{L_b} = \sqrt{D_b \delta}\left(\frac{t}{4D_v}\right)^{1/4}; \tag{2.17}$$

$$\lambda = \frac{\delta}{\sqrt{D_v t}};\ \ \nu = \frac{D_v \sqrt{D_v t}}{D_b \delta}.$$

Following the authors of [74, 78–81], it should be mentioned that in the majority of cases important for practice the dependence $C(x,y)$ or individual characteristics, such as the depth of penetration, the layer concentration curve, are of special interest. Therefore, variables u and w are referred to in [74, 78–81] as dimensionless co-ordinates, the variables λ and ν as the parameters of the dependence realised in every experiment. Partial cases (combination of critical values) of these parameters also give the basic set of asymptotics.

It appears that the claim $D_v/D_b \gg 1$ restricts the possible values of λ and ν. In fact, $\lambda\nu = D_v/D_b$ and consequently, $\lambda\nu \ll 1$. Therefore, the cases $\lambda \gg 1$, $\nu \gg 1$ and $\lambda \sim 1$, $\nu \sim 1$ are not examined. Consequently, we obtain the following set of the regimes of the diffusion process (the names of the regimes were published by the authors of [74, 78–81] in accordance with Harrison's classification):

$$\begin{aligned} &\mathbf{C}: \lambda \gg 1, \nu \ll 1; \\ &\mathbf{B}_1: \lambda \sim 1, \nu \ll 1; \\ &\mathbf{B}_2: \lambda \ll 1, \nu \ll 1; \\ &\mathbf{B}_3: \lambda \ll 1, \nu \sim 1; \\ &\mathbf{B}_4: \lambda \ll 1, \nu \gg 1. \end{aligned} \tag{2.18}$$

It should be mentioned that at fixed temperature T an increase of annealing time t results in a monotonic decrease of the parameter λ whereas ν increases. Thus, the given five stages are presented in the sequence in which they are realised in the experiment with increasing annealing diffusion time at constant temperature or with an increase of temperature at fixed time. The claim according to which the regime of a specific type is realised at the GB indicates that this regime occurs in a large part of the length of the boundary.

The classification (2.18) is not only of the formal nature but has a fully defined physical meaning. For example, parameter λ regulates in equation (2.6) the rations between the value of the derivative of the concentration at the GB in respect of time $\frac{\partial C_b}{\partial t}$ and the term $\left.\frac{2D_v}{\delta}\frac{\partial C_v}{\partial x}\right|_{x=\delta/2}$, responsible for the escape of the diffusant into the volume. At $\lambda \gg 1$ the escape of the diffusant may be ignored. Consequently, (2.6) is written in the new form

$$\frac{\partial C_b}{\partial t} = D_b \frac{\partial^2 C_b}{\partial y^2},$$

which corresponds to the diffusion regime **C** only in the GB. In this case, the solutions have the following form:

$$C(u,v) = C_0 \mathrm{erfc}(u+v) \qquad (2.19)$$

in the Whipple model and

$$C(u,v) = K / \sqrt{\pi D_b t}\, \exp[-(u+v)^2] \qquad (2.20)$$

in the Suzuoka model, where $v = \frac{\omega\sqrt{\lambda}}{2\sqrt{2}} = \frac{y}{2\sqrt{D_b t}}$..

The value of the parameter $\lambda \ll 1$ indicates the quasi-stationary nature of grain boundary diffusion (weak dependence of the concentration of the diffusant at the grain boundary) $\frac{\partial C_b}{\partial t} = 0$, which corresponds to the regime of $\mathbf{B}_2$, $\mathbf{B}_3$ and $\mathbf{B}_4$.

The most general case $\lambda \sim 1$ is realised in the regime **B**.

Parameter ν regulates the relationship between the vertical and horizontal flows of the diffusant in the volume of the specimen. At $\nu \ll 1$ (regimes **C**, $\mathbf{B}_1$ and $\mathbf{B}_2$) the term $\frac{\partial^2 C_v}{\partial y^2}$ in equation (2.5) can be ignored and volume diffusion takes place only in the horizontal direction from the boundary into the body of the grain. At $\nu \sim 1$ and $\nu \gg 1$ (regimes $\mathbf{B}_3$ and $\mathbf{B}_4$ respectively), the vertical flows cannot be ignored and at $\mathbf{B}_4$ they become dominant.

Using this classification, we can obtain asymptotic solutions which are more suitable for processing the experimental data than the complete Whipple and Suzuoka solutions.

2.5. DIFFUSION IN A MOVING GRAIN BOUNDARY. THE CAHN-BALLUFFI DIAGRAM

The phenomenon which often accompanies grain boundary heterodiffusion is diffusion-induced migration of grain boundaries (DIGM). As reported in a review in [82], the grain boundary migration, caused by the effect of a grain boundary diffusion force, was observed for the first time by Raines and Montgomery in 1938. However, its current name DIGM (diffusion induced grain boundary migration) was given to this phenomenon considerably later in a study by Hillert and Purdy [83]. These authors, like Tu [84], are regarded as the people who discovered the examined den Broeder phenomenon [85]. In examining mutual diffusion in a tungsten-chromium binary system, they noted the appearance of migration of tungsten grain boundaries initiated by grain boundary diffusion of chromium. A solid solution of chromium in tungsten formed behind the migration front of the boundaries. This effect could not in principle be generated by volume diffusion from the boundary into the boundary regions of the grain: firstly, because of the low volume diffusion coefficient; secondly the solid solution was found only on one side of the grain boundaries.

In many subsequent studies of DIGM the investigators described the following characteristic phenomenological features of this phenomenon which were generalised in [82, 86].

1. The phenomenon is observed at structurally perfect grain boundaries in efficiently annealed poly- and bicrystals. In this case, the recrystallisation stimulus is small and the boundaries are in the metastable structural equilibrium state.

2. The effect is detected in diffusion pairs with greatly differing values of the partial boundary diffusion coefficients. However, later it was shown [87] that the DIGM phenomenon is also found in diffusion of interstitial impurities.

3. The processes of alloy formation or depletion of the solid solution behind the moving boundary take place at both positive and negative deviations from the ideal mixing energy, i.e. solutions with a positive or negative deviation from Raoult's law.

4. The effect is found at relatively low temperatures when volume diffusion is completely 'frozen' and grain boundary diffusion is extensive. In later studies, DIGM was noted at high temperature when volume diffusion is strong [82].

5. Diffusion-induced grain boundary migration is found only in the case of large-angle grain boundaries and is not found at boundaries with a small misorientation (less than 15°).

6. Boundary migration in the majority of the investigated systems is accompanied by the formation of S-shaped distortions of the initial boundary which grow into adjacent grains.

7. At twin grain boundaries the effect is not found. The curve of the dependence of the distance of migration on the angle of misorientation of the grains shows minima corresponding to special grain boundaries, i.e. the maximum intensity of the effect is found for general type boundaries [88]. However, it has been found that

the DIGM is in fact typical almost any type of GB [89].

The previously mentioned difference between DIGM and conventional recrystallisation was evidently the reason for the absence after initial publications of interest in this phenomenon by experts concerned with the problem of recrystallisation of metals and alloys activated by impurity diffusion [90–95]. It should be mentioned that the phenomenon of activated recrystallisation (AR) was known for a long time and initially reports were published approximately 30 years ago [96, 97]. A series of studies describing the investigation of the relationships and mechanism of the AR process [98–109] were accompanied by the publication in foreign literature of reports on the examination of the effect of diffusion-induced recrystallisation (DIR) [110] in the Fe(Zn) system. It has been established that the DIR process developed like AR by nucleation of new grains in the vicinity of the surface which is in contact with the coating made of another metal, and movement of the DIR front into the thickness of the material. In addition, in some cases there was accelerated nucleation of new grains in the vicinity

Table 2.1. Ratio of the volume diffusion coefficients (D_v) of impurities and rates of BG migration (V_m) in systems with diffusion-induced grain boundary migration, grain boundary sliding, solid state brittleness of metals and structural superplasticity

No.	System	Phenomenon	Homologous temperature	D_v, m²/s	V_m, m/s	D_v/V_m, m
1	2	3	4	5	6	7
1	Ag(Cu)*	DIGM [116]	0.54	–	–	$6 \cdot 10^{-9}$ [116]
2	Ag(Cu)	DIGM [116]	0.63	–	–	$3.3 \cdot 10^{-7}$ [116]
3	Ag(Pd)	DIGM [116,117]	0.33	$9 \cdot 10^{-21}$ [88]	6·10–12	$1.5 \cdot 10^{-9}$ [116]
4	Ag(Pd)	DIGM [116,117]	0.54	–	–	$6 \cdot 10^{-9}$ [116]
5	Ag(Pd)	DIGM [118]	0.63	–	–	10^{-8} [116]
6	Al(Zn)	DIGM [118]	0.47	–	–	$7 \cdot 10^{-7}$ [118]
7	Al(Zn)	DIGM [118]	0.48	–	–	$1.5 \cdot 10^{-6}$ [118]
8	Al(Zn)	DIGM [118]	0.50	–	–	$2.8 \cdot 10^{-6}$ [118]
9	Au(Ag)	DIGM [117, 119–121]	0.40	$5 \cdot 10^{-22}$ [122]	$2 \cdot 10^{-13}$**	$2.5 \cdot 10^{-9}$
10	Au(Ag)	DIGM [117, 119–121]	0.48	$4 \cdot 10^{-19}$ [122]	10^{-10}**	$4 \cdot 10^{-9}$
11	Au(Cu)	DIGM [117]	0.34	$7 \cdot 10^{-22}$ [123]	10^{-12}**	$7 \cdot 10^{-10}$
12	Au(Ni)	DIGM [116]	0.65	–	–	$4.3 \cdot 10^{-9}$ [116]
13	Au(Pd)	DIGM [116]	0.58	–	–	$1.1\ 10^{-7}$ [116]
14	Cu(Al)	DIGM [124]	0.46	10^{-19} [123]	$2 \cdot 10^{-11}$**	$5 \cdot 10^{-9}$
15	Cu(Cd)	DIGM [125]	0.53	–	–	10^{-10}
16	Cu(In)	DIGM [125]	0.53	–	–	10^{-8}
17	Cu(Mn)	DIGM [116]	0,50	–	–	$6 \cdot 10^{-9}$ [116]
18	Cu(Mn)	DIGM [116]	0,57	–	–	$1.3 \cdot 10^{-7}$ [116]
19	Cu(Ni)	DIGM [126]	0.46	$2 \cdot 10^{-22}$	$3 \cdot 10^{-11}$***	$7 \cdot 10^{-12}$

1	2	3	4	5	6	7
20	Cu(Ni)	DIGM [126]	0.68	$1\cdot10^{-19}$	$8\cdot10^{-10***}$	$1.3\cdot10^{-10}$
21	Cu(Ni)	DIGM [126]	0.79	$2\cdot10^{-17}$	$2\cdot10^{-9***}$	10^{-8}
22	Cu(Ni)	DIGM [126]	0.87	$2\cdot10^{-16}$	$4\cdot10^{-9***}$	$5\cdot10^{-8}$
23	Cu(Pd)	DIGM [116]	0.50	–	–	$7.5\cdot10^{-8}$ [116]
24	Cu(Sb)	DIGM [125]	0.53	–	–	10–8
25	Cu(Sn)	DIGM [125]	0.53	$2\cdot10^{-18}$ [96]	10^{-11}–10^{-12**}	$2\cdot10^{-7}$–$2\cdot10^{-6}$
26	Cu(Zn)	DIGM [127]	0.39	$2\cdot10^{-23}$ [90]	10^{-12***}	$2\cdot10^{-11}$
27	Cu(Zn)	DIGM [127]	0.46	$2\cdot10^{-20}$ [90]	10^{-11***}	10^{-9}
28	Cu(Zn)	DIGM [127]	0.53	$2\cdot10^{-18}$ [90]	10^{-10***}	$2\cdot10^{-8}$
29	Cu(Zn)	DIGM [189]	0.51	$1.34\cdot10^{-19}$ [89]	$1{,}28\cdot10^{-10***}$	$1.05\cdot10^{-9}$ [89]
30	Fe(Zn)	DIGM [110]	0.48	–	–	2–$3\cdot10^{-10}$
31	Ni(Cu)	DIGM [126]	0.63	$3\cdot10^{17}$ [128]	$8\cdot10^{-10***}$	$4\cdot10^{-8}$
32	Ni(Cu)	DIGM [126]	0.68	$3\cdot10^{16}$ [128]	$3\cdot10^{-9***}$	10^{-7}
33	Ni(Fe)	DIGM [129]	0.45	10^{-20} [128]	$2\cdot10^{-11**}$	$5\cdot10^{-10}$
34	Pt(Pd)	DIGM [116]	0.38	–	–	$5\cdot10^{-9}$
35	Pt(Pd)	DIGM [116]	0.52	–	–	$2\cdot10^{-8}$
36	W(Cr)	DIGM [117]	0.50	$3\cdot10^{-19}$ [130]	–	$6\cdot10^{-9}$
37	Mo(Ni)	DIR [117]	0.51	$5\cdot10^{-21}$ [130]	$8\cdot10^{-10}$	$0.6\cdot10^{-11}$
38	Mo(Ni)	DIR [131–133]	0.47	$9\cdot10^{-22}$ [130]	–	-
39	Mo(Ni)	DIR [131–133]	0.51	$5\cdot10^{-21}$ [130]	–	–
40	Fe(Ni)	DIGBS [134]	0.5	$1.8\cdot10^{-18}$ [130]	$7\cdot10^{-10}$	$3\cdot10^{-9}$
41	Cu(Bi)	DIGBS [135,136]	0.39	10^{-21} [135]	–	–
42	St.4140 (Sn)	SMB [135]	0.3	10^{-17} [110]	–	–
43	Zn(Hg)	SMB [135]	0.3	10^{-17} [137]	–	–
44	Zn–22 % Al	SSS [138]	0.78	10^{-13} [138]	$3\cdot10^{-7***}$	$3\cdot10^{-7}$
45	Zn–22 % Al	SSS [138]	0.78	–	$4\cdot10^{-8***}$	$3\cdot10^{-5}$
46	Zn–22 % Al	SSS [138]	0.78	–	$4\cdot10^{-9***}$	$3\cdot10^{-5}$
47	MA8	SSS [139]	0.77	$3\cdot10^{-15}$ [139]	$3\cdot10^{-8**}$	10^{-7}
48	12Cr18Ni10Ti	SSS [139]	0.58	10^{-18} [140]	$3\cdot10^{-9**}$	$3\cdot10^{-10}$
49	Steel (Fe–1,6 % C–1,6 % Cr)	SSS [141]	0.51	$2.4\cdot10^{-18}$ [140]	10^{-10**}	$2\cdot10^{-8}$

*The diffusant is given in the brackets. **Migration rate $V_m = X_m/t$, where t is the time during which the boundary migrated to distance X_m. ***Migration rate V_m is given in the cited study of the initial boundaries along which the diffusion flow propagated. The problem of the role of a directional diffusion flow in the development of AR was not discussed in [110].

At present, the processes of diffusion-induced grain boundary migration and recrystallisation have been detected and examined on several pairs of systems (Table 2.1). They include binary single-component systems, alloys, stainless steel, ceramics and calcites. In this case, the impurity diffuses from an external medium which may be represented by a hard metallic coating, liquid or gas.

A large number of attempts have been made to explain the physical natural of the process of DIGM and DIR [82, 83, 111, 146] and construct a phenomenological model of grain boundary diffusion in a migrating grain boundary [65, 81].

The interest in the problems of DIGM and DIR is still strong. Recently, Kajihara and Goust [112] proposed a new model of energy balance for describing the physical nature of DIGM and DIR (the model was constructed on the basis of Gibbs energy diagrams). The quantitative evaluation of the effecting driving force of DIGM and DIR is determined in this model on the basis of the generalisation of the previously available model, Hillert chemical driving force [113] and Hillert and Purdy driving force of coherent stresses [83]. The result is the possibility of evaluating the upper limit of the rate of migration of the GB for the model of coherent stresses $V_m < 10^{-4}D_v/a$ and the lower limit for the model of the chemical driving force $V_m > 8D_v/a$ where V_m is the rate of migration; a is the atomic spacing [112].

Later, Yamamoto and Kajihara [114], using this theory and the data published by Li and Hillert [115] (rate of migration and concentration of the diffusant in the DIGM zone) calculated the effective driving force of DIGM in the Cu(Zn) system at temperatures of 573–773 K. It was found that at T = 573 and 623 K and the maximum rate of migration, the main contribution to this quantity comes from the chemical driving force. In other cases it is necessary to take into account the diffusion of zinc in the non-transformed copper matrix ahead of the moving boundary in accordance with [113]. The temperature dependence of mobility of the GB (ratio of the rate of migration to the effective driving force) shows that DIGM process is controlled by the volume diffusion of zinc in copper [114].

In the latest investigations of the DIGM process using the Cu(Zn) system as an example, the authors of [89] obtained new very important results. They investigated the diffusion of zinc in copper bicrystals where the GB was normal to the [001] axis with different misorientation from 15 to 45°, including special (from $\Sigma5$ to $\Sigma29$) and general GBs. DIGM was found only at T = 693 K and after diffusion annealing for $t = 5.4 \times 10^4$ to 1.73×10^5 s. As expected, after short-term annealing the migration path for general-type grain boundaries was already sufficiently long whereas special-type grain boundaries did not change their positions [89]. This is in good agreement with the previously known facts regarding the effect of the structure of the GB on the kinetics of DIGM [82]. However, at longer annealing times, the grain boundaries of all types migrate and the rate of migration is a constant value [89]. In the quasi-station state of migration the highest rates correspond to special-type grain boundaries and not general ones, as assumed previously in [82]. The authors of [89] explain this by the fact that the grain boundaries were greatly distorted in the DIGM process. As the distance travelled by the boundary increased, the boundary became more distorted. In turn, the curvature of the GB may provide a significant contribution to the energetics of migration reducing the rate of migration as a result of the counteracting change on the form of surface tension. On the other hand, if only the special-type grain boundaries change their position during DIGM, the resultant curvature changes their structure and transforms them into general-type grain boundaries. Thus, this gave a qualitatively new result: the special-type boundaries are characterised by the DIGM phenomenon but this takes place with some time (starting) delay [89]. Evidently, the latter is associated with the build-up of the diffusant at the grain boundary (it is well known that the diffusion permeability of the special-type grain boundaries is lower than that of the general grain boundaries), and also with the need to displace it from the stable equilibrium position corresponding to the low energy state. The experiments show that the chemical driving force is 'balanced' to 75–93% by the volume diffusion of zinc in the initial matrix of copper ahead of the migrating GB.

For the majority of the systems with the DIGM or DIR phenomena, the experimental relationships are explained satisfactorily from the viewpoint of mobility, i.e. variation of the condition of the grain boundaries under the effect of directional grain boundary diffusion flows of impurity [142–146]. In particular, it is assumed that the variation of the state of the boundaries in the examined conditions is the physical reason for the phenomenon, i.e. the diffusion-induced grain boundary sliding [131–134, 143, 144, 146].

The manifestation of the DIRM and DIR effects in many systems (see Table 2.1) in the temperature range corresponding to the development of grain boundary diffusion and also corresponding to the actual surface conditions of structural metallic materials, makes it possible to construct a phenomenological model describing diffusion in migrating grain boundaries adequate to experiments.

The attempts to construct a model taking into account the effect of grain boundary migration on grain boundary diffusion were made in [81]. The Fisher model was used as a basis for making additional assumptions: 1) the boundary moves at constant velocity V_m in the direction of its normal; 2) the diffusant does not interact with the boundary (the boundary is usually separated from diffusing atoms), the concentration of the diffusant has actual values.

Consequently, in the system of co-ordinates linked with the boundary the process is described by the equations (2.5), (2.6) with the difference that the right-hand side contains the additional term $V_m \dfrac{\partial C}{\partial x}$ taking into account the 'flow' of the lattice through the boundary.

The authors of [65, 81] proposed a solution of such a system (Laplace method) for regime $\mathbf{B}_2$:

$$C(u,\omega,\xi) = \frac{2C_0}{\sqrt{\pi}} \exp(-2u\xi) \int_0^\infty \exp(-\alpha^2) \cdot g\left(\frac{\omega^2}{8\alpha^2} + |u|, \xi\right) d\alpha, \tag{2.21}$$

where

$$g(x,y) = \frac{2}{\sqrt{\pi}} \int_x^\infty \exp\left(-\theta^2 - \frac{x^2 y^2}{\theta^2}\right) d\theta.$$

In equation (2.21) ξ denotes the ratio of the displacement of the boundary to the diffusion path in the body of the grain:

$$\xi = \frac{V_m t}{2\sqrt{D_v t}}.$$

It should be mentioned that the quasi-stationary stage in which $\dfrac{\partial C_v}{\partial t} = 0$ and $\dfrac{\partial C_b}{\partial t} = 0$ is interesting. Consequently, the solution of the system is presented in the following form:

$$C_b(y) = C_0 \exp(-y\sqrt{V_m / D_b \delta}), \tag{2.22}$$

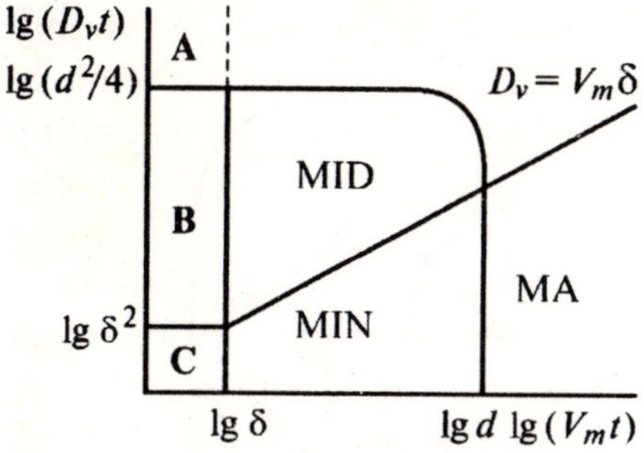

Fig. 2.3. Generalised Cahn–Balluffi classification of diffusion in a migrating GB [145].

$$C_v(x,y) = C_b(y)\exp(-xV_m / D_v),\ x > 0, \tag{2.23}$$

$$C_v(x,y) = C_b(y),\ x < 0. \tag{2.24}$$

The following equation is derived for layer concentration:

$$\overline{C}(y,t) = C_0 V_m t \exp(-y\sqrt{V_m / D_b\delta}). \tag{2.25}$$

The criterion of the start of the stationary state is the value of the dimensionless parameter $\xi \gg 1$, which marginally increases with time. Evidently, as the rate of migration V_m increases, the time to the start of the stationary state decreases. At any $V_m \neq 0$ the stationary stage forms during the time $t \sim D_v / V_m^2$.

It should be mentioned that the described asymptotic solution of (2.21) may characterise only slightly all possible states of the moving GB which have been described extensively in the generalised classification proposed by Cahn and Balluffi [67, 145]. To describe the diffusion process in a migrating grain boundary, it is insufficient to know only the values of $D_v t$ and to characterise the type of diffusion kinetics, the value $V_m t$ must also be used. This value corresponds to the displacement of the boundary during the diffusion experiment t (Fig. 2.3).

At $V_m t < d$ the boundary may be regarded as stationary. In this case we can apply the Harrison classification (see Fig. 2.3 **A**, **B** and **C**). However, at $V_m t > \delta$ the form of the diagram slightly changes. In this case, there are three main kinetic regimes. Zone **MA** corresponds to $(D_v t)^{1/2} + V_m t > d$, where d is the distance between the boundary (or grain diameter). Here diffusion takes place along many boundaries, regardless of whether they move or not (i.e. kinetic regime **A** according to Harrison's classification, partial case **MA**).

Between the lines $(D_v t)^{1/2} + V_m t = d$ and $V_m t = d$ there is a region which corresponds to the case in which the grain boundaries migrate

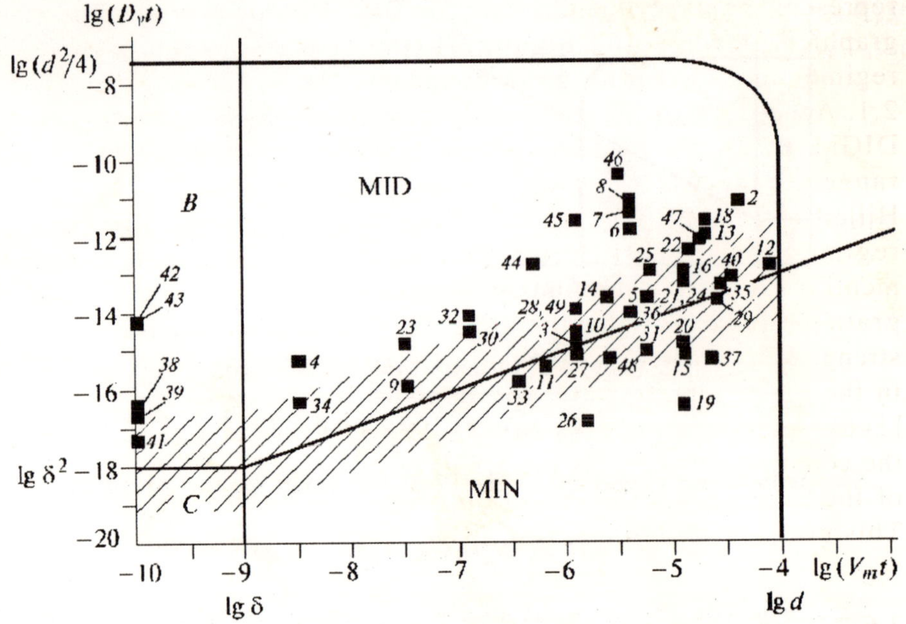

Fig. 2.4. Experimental data for diffusion-induced GB migration represented using the Cahn–Balluffi diagram [145]. The points in the graph correspond to the data in Table 2.1. The separated region is the $\mathbf{B_1}$ grain boundary regime for stationary [74] and 'type $\mathbf{B_1}$' for migrating grain boundaries.

but mass transfer in the lateral direction takes place, because of boundary migration and volume diffusion, over the distance smaller than d and, therefore, the GB may be regarded as isolated. This region is divided into two subzones depending on the intensity of volume diffusion ahead of the migrating boundary. The transition region between these two subzones passes along the line of $V_m t = D_v$ [67] or, by analogy with general classification for stationary boundaries (2.18) [67], it is determined by the value of the dimensionless parameter $\lambda_v = \sqrt{\frac{V_m \delta}{D_v}}$ close to unity. For the case of the migrating boundary this region may be referred to as the regime of 'type $\mathbf{B_1}$' [146]. It may easily be shown that if the boundary overcomes the distance δ during the time $t_b = /V_m$, and the time required by the atom of the diffusant to travel the distance is $t_d = \delta^2/D_v$, and it appears that $t_d \gg t_b$ (or $\lambda_v \ll 1$), the volume diffusion ahead of the moving boundary can be regarded as insignificant (**MIN**). Conversely, at the volume diffusion ahead of the front of the migrating boundary cannot be ignored (**MID**).

Using the Cahn–Baluffi diagram (Fig. 2.3) it appears useful to represent the experimental data for DIGM in an easy-to-understand graphical image [146]. Figure 2.4 (theoretical diagram of diffusion regimes in a migrating grain boundary) shows the data from Table 2.1. According to the previously discussed physical models of the DIGM [112], for systems whose diffusion parameters fall in the range of the **MIN** regime we can use the model proposed by Hillert and Purdy [83]. At diffusion parameters corresponding to the regime **MID**, the Hillert model is suitable [113]. It should be mentioned that the previously mentioned effect of activation of the grain boundaries (the increase of mobility, reduction of shear strength) is detected in the stationary and migrating grain boundaries in the conditions $\mathbf{B}_1$ and 'type $\mathbf{B}_1$ conditions ($D_v = V_m \delta$), respectively [146]. Thus, for the majority of the experimental data in Fig. 2.4 the condition of the possible effect of the directional diffusion flows of the impurity on the condition of the grain boundary is satisfied. This effect will be discussed in greater detail in chapter 4.

2.6. GENERALISED MODELS OF GRAIN BOUNDARY DIFFUSION

the Fisher model is greatly idealised and is not linked directly with the structure of the grain boundary. It should be mentioned that the volume of the grain, even in the case of single-phase single-component materials (pure metals), in the actual experimental conditions is not a homogeneous medium. In all cases, the bi- and polycrystals are characterised by the non-zero density of lattice dislocations. A special case is the mixed ensemble of the grain boundaries which include both high-angle general type and grain boundaries realising the main diffusion mass transfer into the bulk of the specimen, and the low-angle grain boundaries (dislocation walls), and special grain boundaries. The latter, characterised by a considerably higher diffusion permeability in comparisonwith the volume and a considerably lower permability in comparison with general-type grain boundaries, play the role of the routes of intensive outflow of the diffusant from the main direction of diffusion mass transfer. On the other hand, the representation of the grain boundary as a flat homogeneous and isotropic plate with the characteristic width of the order of several interatomic spacings is far from reality and may in some cases lose its descriptive capacity.

However, recently it has become possible to construct generalised models indicating the directions of solving the given problems. For example, attempts have been made to solve the problems of grain boundary diffusion in the presence of a heterogeneity in the core of the grain boundaries [65]. In addition to this, in a general case, the spatially heterogeneous interaction of the atoms of the diffusant with the grain boundaries, leading to segregation, is introduced. The model has been solved for the conditions $\mathbf{B}_1$ and $\mathbf{B}_2$. The main result is the confirmation of the applicability of the standard solutions of the Fisher model to expansion at the heterogeneity of the structure of the core of the grain boundaries with the only difference being that the parameters of the solution become effective quantities characterised by a complicated functional dependence on the microscopic diffusion characteristics of the grain boundaries [65]. Thus, only by greatly simplifying the geometry of the model and introducing 'manually' the structure of the grain boundaries is it possible to obtain quantitative estimates of the macroscopic quantities, whereas the inverse problem of the determination of the microscopic parameters is not solved. In addition to this, the authors of the model [65] do not guarantee that of the resultant equations obtained for the effective diffusion parameters represent a satisfactory approximation of the appropriate physical quantities measured in other conditions of grain boundary diffusion (in different experimental conditions), for example, in the regime **C** [65].

Better results have been obtained by Klinger and Rabkin in taking into account the accelerated path of the outflow of the atoms of the diffusant from the grain boundaries into the volume of the grains, such as the dislocations, sub-boundaries and grain boundaries [147]. The main concept of the authors of the model proposed in [147] is the possibility of avoiding detailed examination of the fine structure of the effective network of the grains, introducing only three qualitatively new parameters: the volume density of the accelerated paths of diffusion in the volume of the grains (defects) g, the coefficient of diffusion along these paths D_d, and the effective coefficient of exchange of the atoms of the diffusant between the nuclei of the defects and the ideal lattice χ. The latter may have a completely determined form for every partial case of the microstructure of the bi- and polycrystals. The qualitative estimates of this parameter may also be obtained on the basis of the generally accepted models. For example, for the outflow of a diffusant along dislocation pipes $\chi = \frac{2D_v}{r_d^2} \ln \frac{r_m}{r_d}$, where r_m and r_d is half the distance between the dislocations and the radius of the dislocation kernel, respectively [147]. In the case of diffusion along the sub-boundaries $\chi = \frac{D_v \pi^2}{\delta(2d_c)}$, where d_c is the size of the subgrains [147]. It is believed [147] that the parameter χ is a phenomenological constant for the selected material and represents a unique gain resulting from avoiding detailed examination of the defective substructure of the grains.

It should be mentioned that a similar concept was proposed independently in [148] as the basis of the model of double diffusion permeability. However, this model is not based on Fisher's representation of diffusion along an isolated grain boundary, and assumes only two types of diffusion path: the ideal lattice and the defect structure–dislocations and grain boundaries [148]. Naturally, this assumption is not sufficient in the case of large differences in the diffusion properties of the dislocations and the grain boundaries.

At the same time, in [147], regardless of the increase of the number of equations in the Fisher model and in the complexity of these equations, it was possible to obtain one of the approximate solutions for the source of constant concentration in the case of quasi-stationary grain boundary diffusion. In [147] it is stressed that the resultant solution has three asymptotics:

1. Short time $\chi t < 1$: the depth of penetration L_b is governed by the Fisher relationship (2.18), but the volume diffusion coefficient D_v is replaced by $g^2 D_d$;
2. Mean times $1 < \chi t < g^{-1}$: the depth of penetration is a weak function of time;
3. Long times $\chi t > g^{-1}$: the depth of penetration along the grain boundary is again governed by the Fisher relationship (2.18), and the volume diffusion coefficient D_v is replaced by the product gD_d.

The first and third of the asymptotic solutions at the given substitutions of the diffusion parameters is efficiently approximated by Le Clare's solution $\ln \bar{c}(y^{6/5})$ with the accuracy to 1.5 and 15%, respectively [147].

On the basis of the resultant asymptotic solutions, the authors of [147] introduced the concept of the three new conditions of grain boundary diffusion – $\mathbf{D}_1$, $\mathbf{D}_2$, and $\mathbf{D}_3$ which do not differ qualitatively from the well-known classification (2.18) and correspond to the three diffusion times through sub-structural defects. The sequence of the regimes in the diffusion experiment depends on the ratio of the values D_v and g^2D_d, gD_d. For example, if $D_v>gD_d$, then the sequence of conditions (2.18) applies in the experiment. At $D_v<g^2D_d$, the regime **C** transfers to $\mathbf{D}_1$, $\mathbf{D}_2$ and $\mathbf{D}_3$, and the conditions of type **B** are not realised. In the intermediate case $g^2D_d<D_v<gD_d$, all the regimes **B** and **D** are found. In a general case, their sequence has the form: $\mathbf{C}\rightarrow\mathbf{D}_1\rightarrow\mathbf{B}\rightarrow\mathbf{D}_2\rightarrow\mathbf{D}_3$ [147].

On the basis of the proposed classification [147] it has been possible to explain several low-temperature anomalies of grain boundary diffusion [149], introducing the critical volume fraction of the defects $g^*=\dfrac{s\delta}{2\sqrt{D_d t}}$. If the diffusion experiment is carried out in the conditions of expectation of the grain boundary regime **C** according to the available classification (2.18), and the volume fraction of the defects is $g > g^*$, the observed profile of the concentration has the form corresponding to the Whipple–LeClair equation (2.16), with the variation $D_v \rightarrow g^2D_d$ introduced in accordance with the representations regarding the regime $\mathbf{D}_1$ [147].

Thus, regardless of the qualitative possibility of the expanded Fisher model of grain boundary diffusion to solve the problem of the sequence of the diffusion regimes for the case of an inhomogeneous structure of the grains, the qualitative analysis may prove to be greatly complicated because of the difficulties in the determination of the value of g, i.e. the volume fraction of substructural defects. In particular, this relates to the previously mentioned case of the mixed ensemble of the grain boundaries detected in nanostructured materials [64], produced in particular by the effect of severe plastic deformation [150].

It should be mentioned that the classification of the conditions of grain boundary diffusion for the metals and alloys with small grain sizes requires clarification proposed by Mishin and Herzig in [151]. For example, it is proposed to formally name the regimes **C**, $\mathbf{B}_1$ and $\mathbf{B}_2$ as **C'**, $\mathbf{B}'_1$ and $\mathbf{B}'_2$, respectively, if the depth of penetration of the diffusant $L_b = (D'_b)^{1/2}$ for the regime **C** and $L_b=\dfrac{(s\delta D_b)^{1/2}}{(4D_v/t)^{1/4}}$ (2.17) for the type **B** regimes exceeds the grain size d [151]. The regimes **A** and **A'** differ greatly [65]. As mentioned previously, the regime **A** corresponds to long annealing times (high annealing temperatures) when the diffusion flows from the adjacent grain boundaries overlap. However, for the coarse-grained materials it is

Table 2.2. Diffusion regimes in polycrystals of different classes [65]

Polycrystal class	Sequence of realisation of regimes in time
Coarse-grained	**CB_2-B_4-A**
Fine-grained	**C-B_2-B'_2-A'**
Ultrafine-grained	**C-C'-B'_2-A'**
Special case of ultrafine-grained	**C-C'-A_0**

assumed that the volume diffusion directly from the surface has a very strong (even dominant) effect in this case. Consequently, the depth of penetration, measured in the experiments, depends on the volume diffusion coefficient [67]. Another case is the regime **A'** when the grain size d is of the order of the diffusion length in the volume $L = (D_v t)^{1/2}$. At the same time, the depth of penetration L_b along the grain boundaries greatly exceeds both d and L. In this case, direct volume diffusion may be ignored, and the dependence of the layer concentration on the depth has the following form [65]:

$$\bar{c} = c_0 \operatorname{erfc}\left[\frac{y}{2(D_{eff}t)^{1/2}}\right], \tag{2.26}$$

where $D_{eff} = \zeta s\delta D_b/d$; $\zeta = 2.932$ is the geometrical factor. It is stressed [65] that in the regime **A'** it is possible to determine independently the grain boundary diffusion coefficient D_b.

Depending on the sequence of realisation of the diffusion regimes, in [65] the materials are divided formally into diffusion classes (Table 2.2). A special class of ultrafine-grained polycrystals has been determined on the condition $s\delta/2 > d$ (regime $\mathbf{A}_0$). Consequently, the effective diffusion coefficient is calculated from the equation (2.3), where the expression for the volume fraction of the grain boundaries is re-written in the following form [65]:

$$\tau = \frac{\zeta s\delta/d}{1+\zeta s\delta/d}. \tag{2.27}$$

In regime $\mathbf{A}_0$ it is assumed that $s\delta/2 > d$ and, consequently, $\tau \sim 1$, $D_{eff} = D_b$. Consequently, the concentration profile corresponds to the equation (2.26), where $D_{eff} = D_b$ [65].

Thus, the Fisher model of grain boundary diffusion, constructed more than 40 years ago, is still the basis for the construction of new improved models and processing of the experimental data of investigations of self- and heterodiffusion along the single misorientation boundaries in the bicrystals and grain boundaries in the polycrystals.

A classification of the conditions of grain boundary diffusion in the polycrystals and bicrystals has been proposed, models of diffusion in a moving grain boundary and methods of constructing diagrams, reflecting the pattern of development of the diffusion process in polycrystals with migrating grain boundaries have been constructed. The phenomena of diffusion-induced migration of grain boundaries, recrystallisation and grain boundary sliding, discovered in the last decade and associated with the grain boundary diffusion of impurities, have been examined in detail.

The main relationships of grain boundary diffusion have been examined qualitatively, and the conditions of grain boundary diffusion for the case of the small grain size and the internal inhomogeneity of the grains have been classified. Each of the conditions is characterised by the well-known type of concentration curves, has a specific position in the sequence of realisation of the diffusion regimes in the experiments and, at the known parameters of structural special features of the material, makes it possible to calculate, using the experimental data, the relationships of the main diffusion parameters and, in some cases, the exact values of these parameters.At the same time, a general model of the diffusion process in the nanostructured and nanocrystalline materials has not as yet been completely developed because of large differences in the structure of these materials produced by different methods. In addition to this, it is believed that such a model will not be developed in the near future echoes the introduction of additional parameters complicates even more appreciably the diffusion model and of the application of its solutions for describing the experimental data.

CHAPTER 3

SPECIAL FEATURES OF GRAIN BOUNDARY DIFFUSION IN NANOCRYSTALLINE AND NANOSTRUCTURED MATERIALS

In this chapter, attention is given to the literature and original results of experimental investigations of the diffusion process in nanocrystalline and nanostructured materials. Special attention is given to insufficiently examined special features of grain boundary diffusion in bulk NS metals produced by the effect of severe plastic deformation. These metals are promising for extensive application in technology, including as structural materials.

3.1. SPECIAL FEATURES OF GRAIN BOUNDARY DIFFUSION IN NANOCRYSTALLINE MATERIALS

Gleiter [64] reviewed the literature data available at that time obtained in the initial investigations of diffusion in NC materials. In particular, it is shown that the initial investigation of the special features of self-diffusion in NC metals was carried out on the example of nanocrystalline copper (d~10 nm) produced by condensation in an inert gas followed by pressing under pressure. After depositing a thin layer of ^{67}Cu isotope and diffusion annealing at 293, 353 and 393 K by ion sputtering it was possible to construct concentration profiles. According to the author of the review [64], the conditions of realisation of regime C were fulfilled in the experiment and, using the equation for an instantaneously exhausted source (2.20), true values of the self-diffusion coefficient D_n in NC copper were determined. These values were several orders of

Table 3.1. Diffusion coefficients in nanocrystalline and nanostructured materials

Material	d, Nm	Diffusant	T, K	Method	D, M^2 s			Source
					D_n	D_b	D_v	
NC Cu	10	^{67}Cu	293	RA	$2{,}6 \cdot 10^{-20}$	$4, \cdot 10^{-24}$	$4{,}0 \cdot 10^{-40}$	Horvath et al. [152]
	10	Ag	293	XSMA	$4{,}8 \cdot 10^{-17}$	$8{,}1 \cdot 10^{-20}$	$8, \cdot 10^{-33}$	Schumacher et al. [152]
	10	Au	293	AES	$4{,}7 \cdot 10^{-22}$	$2{,}6 \cdot 10^{-26}$	$1{,}6 \cdot 10^{-34}$	Hoefler [152]
	10	Bi	293	RBS	$2{,}3 \cdot 10^{-19}$	–	$8{,}8 \cdot 10^{-30}$	"
NC Pd	8–11	Ag	393	SIMS	$3{,}5 \cdot 10^{-17}$	$1{,}3 \cdot 10^{-21}$	$5.8 \cdot 10^{-25}$	"
	8–11	Au	393	SIMS	$8{,}1 \cdot 10^{-19}$	$1{,}5 \cdot 10^{-20}$	$1{,}3 \cdot 10^{-29}$	"
	8–11	Cu	393	SIMS	$1{,}1 \cdot 10^{-18}$	–	$2{,}1 \cdot 10^{-36}$	"
	8–11	B	393	SIMS	$3{,}0 \cdot 10^{-24}$	$3{,}7 \cdot 10^{-22}$	–	"
	8–11	O	393	SIMS	$8.5 \cdot 10^{-18}$	–	–	"
	5	H	293	RBS	$5{,}1 \cdot 10^{-12}$	–	$6{,}1 \cdot 10^{-11}$	Mutschele et al. [152]
NC Fe	7	B	298	SIMS	$3{,}4 \cdot 10^{-21}$	–	$2{,}5 \cdot 10^{-20}$	Hoefler et al. [152]
NC TiO_2	20–25	Hf	920	RBS	$3.0 \cdot 10^{-15}$	–	$3{,}0 \cdot 10^{-16}$	"
	20–25	^{18}O	920	SIMS	$4.0 \cdot 10^{-12}$	–	$5{,}0 \cdot 10^{-18}$	"
NC Ni	70	Au	448	RBS	$3{,}3 \cdot 10^{-20}$	$5{,}6 \cdot 10^{-21}$	$3{,}1 \cdot 10^{-35}$	Bokshtein et al [156]
	70	^{63}Ni	293	RA	$1{,}14 \cdot 10^{-20}$	$1{,}4 \cdot 10^{-26}$	$3{,}4 \cdot 10^{-54}$	"
	70	^{63}Ni	373	RA	$1{,}05 \cdot 10^{-18}$	$3{,}3 \cdot 10^{-22}$	$1{,}2 \cdot 10^{-43}$	"
	70	^{63}Ni	423	RA	$4{,}0 \cdot 10^{-18}$	$2{,}7 \cdot 10^{-20}$	$4{,}3 \cdot 10^{-39}$	"
	70	^{63}Ni	473	RA	$2{,}0 \cdot 10^{-17}$	$8{,}4 \cdot 10^{-19}$	$1{,}7 \cdot 10^{-35}$	"
NC Ni	30	Cu	423	SIMS	$3{,}8 \cdot 10^{-17}$	$4{,}31 \cdot 10^{-19}$	–	Kolobov et al [157]
NC Pd	20	^{59}Fe	473	RA	$4{,}0 \cdot 10^{-20}$	$1.3 \cdot 10^{-21}$	–	Schaefer et al. [161]
	52	^{59}Fe	523	RA	$1{,}0 \cdot 10^{-19}$	$1{,}0 \cdot 10^{-19}$	–	"
NC Fe	31	^{59}Fe	452	RA	$2{,}4 \cdot 10^{-21}$	$1{,}6 \cdot 10^{-22}$	–	Schaefer et al. [163]
	31	^{59}Fe	472	RA	$9{,}4 \cdot 10^{-21}$	$1{,}3 \cdot 10^{-21}$	–	"
	31	^{59}Fe	499	RA	$2{,}9 \cdot 10^{-20}$	$2{,}9 \cdot 10^{-20}$	–	"
NS Pd	80–150	^{59}Fe	371	RA	$1{,}0 \cdot 10^{-21}$	$8{,}0 \cdot 10^{-25}$	–	Schaefer et al. [161]
	80–150	^{59}Fe	401	RA	$3{,}0 \cdot 10^{-20}$	$1{,}2 \cdot 10^{-23}$	–	"
	80–150	^{59}Fe	473	RA	$5{,}0 \cdot 10^{-20}$	$6{,}0 \cdot 10^{-21}$	–	"
	80–150	^{59}Fe	577	RA	$9{,}0 \cdot 10^{-20}$	$3{,}0 \cdot 10^{-18}$	–	"
NS Fe	150	^{59}Fe	528	RA	$4{,}0 \cdot 10^{-19}$	$1{,}0 \cdot 10^{-19}$	–	"
NS Ni	300	Cu	398	SIMS	$5{,}06 \cdot 10^{-15}$	$4{,}64 \cdot 10^{-20}$	–	Kolobov et al [157]
	300	Cu	423	SIMS	$9{,}6 \cdot 10^{-15}$	$4{,}31 \cdot 10^{-19}$	–	"
	300	Cu	448	SIMS	$2{,}2 \cdot 10^{-14}$	$2{,}14 \cdot 10^{-18}$	–	"

Diffusion coefficients: D_n – in nanocrystalline and nanostructured materials, D_b – grain boundaries, D_v – in the volume of coarse-grained materials. RIA – Radio Isotope Analysis, MXSA – X-ray Spectrum Analysis, AES – Auger Electron Spectroscopy, RBS – Rutherford Backscattering, SIMS – Secondary Ion Mass Spectroscopy, EDM – Measurements of Electromotive Force.

magnitude higher than the coefficients of both volume D_v and grain boundary diffusion D_b in polycrystalline copper (Table 3.1). The temperature dependence of D_n was used to determine the activation energy of self-diffusion $Q^n_b = 61.8$ kJ/mol. The latter is considerably lower than the appropriate value for polycrystals ($Q_b = 102$ kJ/mol) and single crystals ($Q_1 = 191$ kJ/mol) of copper.

In the review [64] special features of heterodiffusion of silver in NC copper were investigated. The concentration profiles of silver were constructed on the basis of the results of x-ray micro spectrum analysis. It was found that below 343 K in NC copper the activation energy of diffusion of Ag is $Q^n_b = 65.6$ kJ/mol. The temperature dependence $D_n(T)$ of diffusion of silver in nanocrystalline palladium in the temperature range 293–453 K was also investigated [64]. The resultant values of the activation energy of grain boundary diffusion $Q^n_b = 42.5$ kJ/mol was approximately half the activation energy of grain boundary diffusion in polycrystalline palladium [64].

Gleiter [64] assumed the existence in NC materials of an 'amorphous' grain boundary phase which in the investigated material represents up to 50% of the general mass and is characterised by the absence of not only long-range but also short-range order in the distribution of atoms. This contradicts the experimental data indicating that the rate of diffusion of substitutional impurities in amorphous metallic alloys is considerably lower than the appropriate rate for grain boundary diffusion in polycrystals [152]. At the same time, the concept of the 'amorph' was transformed recently into a new concept of 'gel-like' state of GB in gas-condensed NC materials formed on the basis of molecular-dynamic modelling [153].

The conclusion of the author of [64] regarding the considerably higher values of true coefficients of grain boundary diffusion in NC materials was criticised by Klotsman [154]. Klotsman assumed that in [64] the experimental data were not interpreted using a diffusion model adequate for the given experiments. In particular, the assumption made in [64] on the realization of the regime **C** in the investigative conditions does not correspond to reality and there is a large outflow of the isotopes ^{67}Cu from the grain boundary to the boundary volumes. In this case, the volume diffusion coefficient in the boundary zone (D'_v) greatly differs from the appropriate value in the single crystal. Because of the small grain size ($d \sim 10$ nm) diffusion flows from the GB may overlap within the limits of the same crystal leading to the situation similar to diffusion in a quasi-homogeneous specimen (regime **A** according to Harrison's classification). On the other hand, the previously mentioned review of application of Fisher's models [65, 67, 74, 78–81, 147] makes it possible to claim unambiguously that of all the possible solutions, corresponding to the experimentally observed dependence of layer concentration on depth ln $\bar{c}(y^2)$ the lowest value of the true coefficient of grain boundary diffusion is found in regime **C** [65] (compare, for example, equations (2.19) and (2.26)); the calculations of the diffusion parameters in [64] were also carried out assuming that this regime does occur.

An interesting continuation of this kind of considerations are the experimental data obtained for the diffusion of boron in NC iron [155]. This type of diffusion belongs, according to the authors of [155], to the regime similar to regime **C** but for the case $D_v > D_b$. Diffusion takes place only along the boundary because the solubility of the diffusant in the crystal lattice is very low (in diffusion of boron in iron the coefficient of grain boundaries segregation is $s = 10^4$). Consequently, $D_{eff} = 1.7 \cdot 10^{-11} \exp(-556.9/RT)$ m²/s and it is possible to compare it with $D_v = 2.3 \cdot 10^{-7} \exp(-79.1/RT)$ m²/s for diffusion of boron in iron single crystals.

It should be mentioned that the possibility of evaluating the minimum value of the coefficient of grain boundary diffusion in the approximation of realization of regime **C** can also be extended to the case of quasi-stationary diffusion (regimes **B** and **D**). Even in cases in which the exact form of the curve ln $\bar{c}(y)$ cannot be determined because of the experimental error, it appears to be sufficient to separate on this curve the linear section with the minimum angle of inclination $\frac{\partial}{\partial y} \ln \bar{c}(y)$, extrapolate the value c_0 and, substituting it into the equation for layer concentration in the regime **C** (2.19), the experimental value of the latter at the maximum depth y corresponding to the end of the linear section, to calculate the diffusion coefficient D which is a lower estimate of the true grain boundary diffusion coefficient D_b.

Later, Bokshteln in et al [156] examined in detail self-diffusion in NC nickel produced by the same method [64] (condensation in an inert gas followed by pressing under pressure at $T = 773$ K to a density of 92–93%). The grain size of this material was 70 nm. The results of examination of diffusion in annealing in the temperature range 293–473 K (see Table 3.1) were used to conclude that grain boundary diffusion in regime **C** changes to diffusion in regime **B** (according to Harrison's classification) at

T = 473 K (6 h). As assumed in [154], the coefficient of diffusion of marked nickel atoms from the grain boundary into the boundary volume is considerably lower than the coefficient of volume self-diffusion in single crystal nickel. In addition to this, at T = 473 K $D'_v/D_v \sim 10^{16}$. To verify this result, the heterodiffusion of gold in NC Ni at T = 448 K was investigated. As in the case of self-diffusion, it was shown that $D'_v/D_v \sim 10^{12}$ for the heterodiffusion of gold in nickel at this temperature. However, in analysis of the experimental data [156] the claim on the large increase of the coefficients of grain boundary diffusion in NC metals is valid (see Table 3.1). In the cited study it was also found that for the grain boundary self-diffusion of NC nickel Q^n_b = 46 kJ/mol, whereas the activation energy of grain boundary self-diffusion in polycrystalline nickel is Q_b = 115 kJ/mol. The authors of [156] show that the value 46 kJ/mol is close to the activation energy of self-diffusion on the free surface of nickel.

The results presented in [156] were used to support the cluster model of the structure of NC materials. It is assumed that nanocrystals (d~100 nm) form clusters with the size of 1–10 μm. As a result, the grain boundary ensemble in this case contains GBs of two types: GBs inside the cluster which do not differ from conventional GBs in polycrystals, and intercluster boundaries with the concentration of the entire excess volume of the NC materials. In this model, D_n characterises diffusion along intercluster boundaries, and D'_v is synonymous with the effective coefficient of outflow of the diffusant from the intercluster boundaries to the volume of the cluster along the GB inside the cluster with the diffusion coefficient D_b characteristic of polycrystals, but taking into account the density of 'sinks' (intercrystalline boundaries) $a \approx \pi\delta/2d \sim 0.022$ [156]. The latter claim is in relatively good agreement with experiments. It may be said that the authors of [156] confirmed by experiment the possibility of realization of regime $\mathbf{D}_1$ in examining the diffusion in NC metals. Taking the current views into account, we can make a correction according to which the value D'_v should not be regarded as synonymous with $D_b \cdot \alpha$, but with the product $D_b \cdot \alpha^2$ according to the definition of regime $\mathbf{D}_1$. On the other hand, this comment has no significant effect on the qualitative and quantitative agreement between the experiments [156] and theory [147].

In [157] the authors determined the coefficients of grain boundary diffusion of copper in NC nickel. It is shown that in the process of diffusion annealing at 423 K there is no grain growth and the estimate of the upper value of the diffusion coefficient of $3.8 \cdot 10^{-17}$ m^2/s is two orders of magnitude higher than the calculated value for the diffusion of copper in coarse–grain nickel [157]. In addition to this, this value is in almost complete agreement with the value obtained for the coefficient of self-diffusion of nickel for the same temperature published in [156] (Table 3.1).

Larikov [152] shows that the analysis of the available data does not provide sufficient grounds for giving the empirical values of Q''_b the meaning of the activation enthalpy of an elementary physical process, and these values are only 'effective temperature coefficients of the rate of mass transfer' in NC materials. He assumes that 'if the presence of a small number of isolated micropores can hardly influence macroscopic diffusion, then sub-micropores (sources and sinks for vacancies) and vacancies themselves play a significant role in accelerating diffusion in NC materials'. This explanation of the high diffusion coefficients is in good agreement with the cluster model of NC materials, proposed in [156], which has been confirmed in [158] in examining, by high-resolution electron microscopy, specimens of silver and palladium (d~8–10 nm) prepared by the same technology [64]. Pores were found at triple joints at the grain boundaries in these systems. The presence of the amorphous or spherical 'grain boundary' phase, broadening of the grain boundaries or distortion of the crystal lattice in the boundary regions, were not found. The confirmation of the above views regarding the structural state (in particular, the absence of the amorphous phase) of the NC materials produced by deposition in an inert gas, was made in [159]. Examination by the method of inelastic neutron scattering on the example of NC nickel showed that the spectrum of up to 12% of atoms in this material changes. The changes observed here correspond to a two-dimensional system, i.e. a spectrum corresponding to the state of atoms on the free surface which, however, is not found in the case of non-compacted nickel nanopowders. Consequently, the authors of [159] concluded that the state (vibrational behaviour) of the grain boundaries in NC metals is special.

By analogy with the structure of the material (NC palladium) the authors of [160] examined the fine structure of the spectra of absorption of x-radiation (EXAFS spectroscopy). They determined the functions of the radial distribution of the atomic density for nano- and polycrystalline palladium and also coordination numbers and atomic spacings. The results show that the NC specimens are crystallographically identical to the poly- and single crystals of palladium. The presence of non-ordered phases, relating to the grain boundaries and the space between the boundaries, was not detected. The observed decrease of the coordination numbers is explained by the authors of [160] by the small grain size and vacancy disordering of the crystal lattice of NC palladium with the concentration of lattice vacancies of 2–3%. This is confirmed by the results published in [161] where analysis was made of the spectra of the lifetime of positrons in NC palladium and, consequently, distribution of the free volume. The experiments show that gas–condensed Pd contains both pores and sub-micropores and also a high concentration of single vacancies [161]. At the same time, in [161] examination of diffusion in densely compacted (98% of theoretical density) NC palladium shows that the coefficients of grain boundary diffusion of ^{59}Fe and ^{103}Pd isotopes do not differ from the corresponding calculated values for the coarse-grain materials (Table 3.1). The same result (Table 3.1) was obtained by Tanimoto et al [162, 163] in examining self-diffusion in nano-crystalline iron in the densely compacted state, including the state produced by explosion pressing [163]. However, the main factor which may influence the diffusion properties is the migration of the GBs detected in all experiments [161–163].

3.2. EXAMINATION OF GRAIN BOUNDARY DIFFUSION IN NANOSTRUCTURED METALS PRODUCED BY SEVERE PLASTIC DEFORMATION

The discussion of the problems of the effect of the small grain size and, consequently, overlapping of the diffusion flows of impurity atoms from the grain boundary to the boundary volumes of the grains, and also the effect of residual porosity of NC materials on the course of diffusion processes has been continued in investigations of the processes of grain boundary heterodiffusion in MS materials with similar structure and properties (grain size d~0.1–0.3 μm), produced by severe plastic deformation. One of these methods is equal–channel angular pressing (ECAP) described in Chapter 1. The materials produced by ECA pressing do not contain pores and can be used to produce specimens with relatively large dimensions. The state of the GBs in these materials is characterised by high non-equilibrium (presence of high elastic stresses, intrinsic and lattice dislocations, facets and steps of deformation origin on the GBs). The boundary regions are characterised by large distortion of the crystal lattice, directly indicating the presence of elastic stresses and, consequently, the non-equilibrium state of the GBs [12, 150, 164]. More details on the mechanical and other properties of the NS materials will be presented in the following chapters.

Diffusion in these materials was investigated for the first time by Wurschum, Schaefer, et al. [161, 165, 166] on the example of diffusion of isotopes of ^{59}Fe in Pd. They did not find any large increase of the grain boundary diffusion coefficient (Table 3.1). Only at the lowest test temperatures (T = 371.401 K) the difference is not greater than two orders of magnitude [161]. At the same time, the activation energy of grain boundary diffusion in NS palladium was 1.5 times lower than in coarse-grained palladium [165]. It should be mentioned that migration of the grain boundaries was found during diffusion annealing [161] and influenced not only the model used for estimating the diffusion coefficients but also changed the natural state of the GBs which, as mentioned in the above studies, is highly non-equilibrium. Thus, the recovery in the structure of the grain boundaries in NS palladium during diffusion annealing may greatly influence the diffusion properties of the materials [161]. Similar results were obtained later in examining self-diffusion in NS iron [162. 163] where no increase of the coefficient was again found (Table 3.1).

Examination of the diffusion parameters of the NS materials was carried out in [167–170] on the example of diffusion of copper in NS nickel. The concentration of copper in nickel at various depths was measured by the method of secondary ion mass spectroscopy (SIMS). For comparison, the authors of [168] used original data for the diffusion of copper in coarse–grain nickel in the temperature range 773–873 K. It was found that after annealing at temperatures below 573 K copper is not found in coarse–grain nickel even at a depth of 1–2 μm. Only after annealing at T = 773, 823 and 873 K was it possible to construct concen-tration profiles (Fig.3.1, curves *1–3*) and calculate the coefficients of grain boundary diffusion of

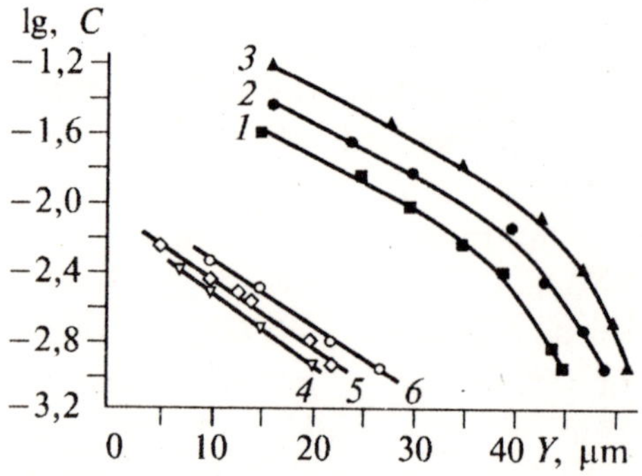

Fig.3.1. Profiles of concentration of copper in coarse grain (curves *1–3*) and nano-structured (curves *4–6*) nickel after diffusion annealing at T = 773 (*1*) 823 (2) 873 (*3*) 398 (*4*) 323 (5) 448 K (*6*) [168].

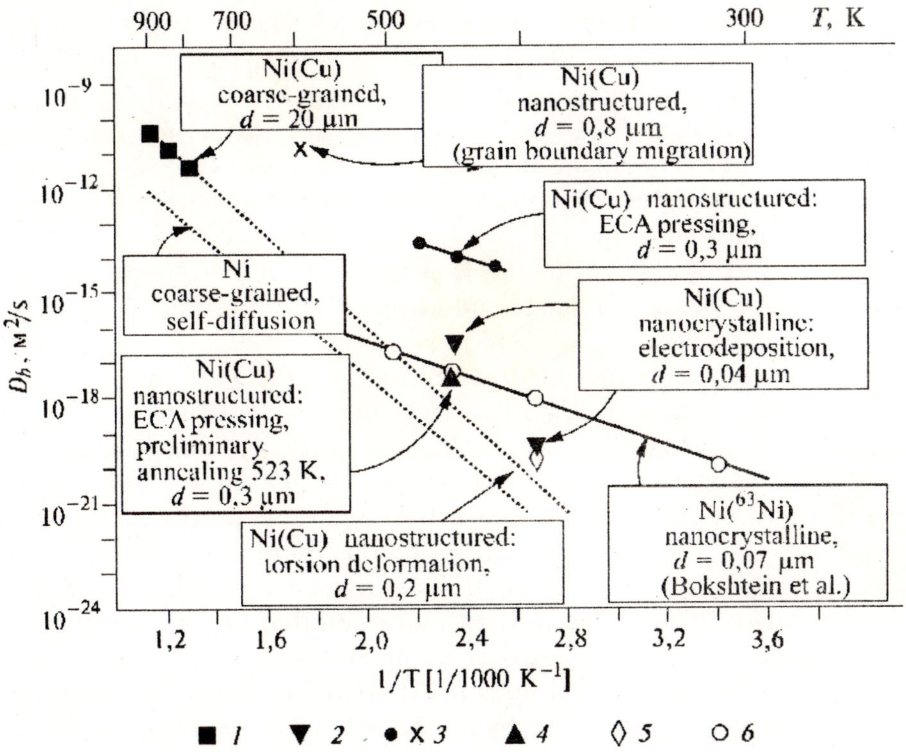

Fig.3.2. Temperature dependence of the coefficients of diffusion of copper in nickel with different structures: in coarse-grain *(1)* [157]; Nanocrystalline produced by electrodeposition *(2)* [157]; in nanostructured, produced by equal-channel angular pressing, including: in the initial condition (and after preliminary annealing at T = 523 K for 1 h *(4)* [157]; in the nanostructured condition, produced by torsional deformation by torsion *(5)*. For comparison, the currently available experimental data for self-diffusion in nanocrystalline nickel *(6)*, [156] are also given.

copper in coarse-grain nickel. It was taken into account [168] that in the annealing conditions, the rates of migration of the grain boundary measured in experiments at temperatures of 773, 823 and 873 K were $1.5 \cdot 10^{-11}$, $0.6 \cdot 10^{-10}$ and $1.9 \cdot 10^{-10}$ m/s, respectively. In this case, the diffusion coefficient maybe calculated using equation (2.25) for migrating boundaries. The activation energy of the grain boundary diffusion of copper in coarse-grain nickel was Q_b = 124.4 kJ/mol (Fig.3.2) [168].

Assuming that the parameters of the Arrhenius dependence for the grain boundary diffusion coefficients are constant in the entire investigated temperature range, the values of D_b were calculated for coarse-grain nickel at lower temperatures [168].

Figure 3.1 (curves 4–6) gives the concentration profiles of the distribution of copper and the thickness in NS nickel after annealing in the temperature range 398–448 K for 3 hours [168]. Taking into account that the grain boundaries in the NS nickel do not migrate in the temperature range 392–448 K (according to transmission electron-microscopy data [168]), the diffusion coefficients (Table 3.1) were calculated in the approximation of realization in the investigated annealing conditions of the regime of grain boundary diffusion C using equation (2.19). As shown by Table 3.1, the values of copper in NS nickel are 4–5 orders of magnitude higher than the corresponding values in coarse-grain nickel [168] .

The knowledge of the special features of the structures of NS nickel can be used to evaluate the grain boundary diffusion regimes realized in the diffusion experiment. As already mentioned, the widening of the GB, distortion of the lattice in the boundary regions, and, consequently, possible changes to the diffusion parameters [65] do not cause any change in Fisher's model and they result only in the shift on the sequence of realization of the effective diffusion regimes [65]. A more detailed discussion of this problem requires more accurate experimental data, on the one hand, and theoretical considerations on diffusion in fields with steep stress gradients, on the other hand. At this time, investigations of this type cannot be carried out. At the same time, we believe that there is sufficient amount of information for constructing a sequence of the grain boundary regimes for the nickel–copper system realized in the experiment. For example, according to the classification [151], the NS nickel fits in the range of ultra-fine grains polycrystals (Table 2.2). This means that the sequence of the regimes $\mathbf{C}\rightarrow\mathbf{C}'\rightarrow\mathbf{B}'_2$ is realized in this material. In addition to this, regime $\mathbf{D}_1$ can be realized [147] (or $\mathbf{D}'_1$ by analogy with the regime $\mathbf{C}'$ and $\mathbf{B}'_2$ for diffusion through the network of the grain boundaries). For a given dislocation density in NS nickel ($\rho \sim 7\cdot10^{14}$ m^{-2}) it appears that $g \leq 10^{-4}$. Taking into account the high volume fraction of the sub boundaries we can obtain the upper estimate of g. It is assumed that at the distance d (grain size) from each other the grain boundaries intersect sub-boundaries. This means modelling a mixed ensemble consisting of low-angle, special and high-angle boundaries. It is also assumed that the network of the high-angle grain boundaries penetrate through the entire specimens (the possibility of percolation for the diffusion flow) and provides the main contribution to diffusion mass transfer. This results in the volume fraction of defects of $g{\sim}d/d{\sim}10^{-3}$. Assuming that the dislocation mechanism of diffusion outflows along the sub-boundaries operates ($D_d \sim 10^{-21}$–10^{-19} m²/s at T = 398–448 K), it maybe shown that the duration of transition of the regimes **C→D¹** (or **C→D'**$_1$) $t = \delta^2/(4D_d \cdot g^2)$ from ~10³ h at T = 398 K to ~6 h at T = 448 K. Thus, in actual experiments we may observe the transition from diffusion in regime **C'** to diffusion in regime **D'**$_1$.

It should be mentioned that the network of defects with the possibility of percolation for the diffusion flow in a polycrystalline material is also produced by triple joints (TJ), in addition to the grain boundaries. The diffusion parameters of the TJ may greatly differ from those of the GB. Evidently, these defects require special attention when discussing diffusion in MS materials where the volume fraction of the TJ is considerably higher [171]. If the main diffusion mass transfer takes place along the TJ, we can again use Fischer's model for the TJ utilizing the hierarchy of the defects, as carried out in [147]. Assuming that the efficiency of TJ → GB interaction is considerably higher than the diffusion exchange of GB and the defective boundary volume of the grains, we can construct a model in which the effective parameters are the diffusion coefficients along TJ (D_{TL}), the diffusion coefficient characterising the outflow of the impurity from TJ along GB (D_b) (when solving the problem of diffusion along TJ it is not necessary to transfer to cylindrical coordinates because the outflow of the diffusant TJ → GB takes place in the plane of location of the GB, which corresponds to the two-dimensional Fischer model). Thus, the volume fraction of 'defects' (GB) (g) is of the order of unity. Thus, the classification proposed in [147] can also be used for diffusion along the TJ. It appears that the sequence of the diffusion regimes has the transition **C'→D'**$_1$ in the very early stages of diffusion annealing (~1 s if we use the values of D_b, calculated in [168]), because in this case it may be concluded that regime **C** does not form here. Consideration of the GB–defect of the grain exchange influences in all likelihood and has only a slight effect during the diffusion and annealing period, as mentioned in examining the model of the diffusion process with dominant mass transfer along the GB However, a much stronger role by the possible interaction (intersection) of diffusion flow along the grain boundaries from the adjacent triple joints results in a situation qualitatively identical with the formation of regime **A'** in accordance with the classification [151] for ultrafine grain polycrystals. Here the concentration profile is described by the equation (2.26) where $D_{eff} = \varsigma F D_b$ or $D_{eff} = \varsigma F_{TL} D_{TL}$ for diffusion along a triple joint; F is the ratio of the volume fraction of the continuous network of the defects which carries out main diffusion mass transfer, to the volume fraction of the network of defects through which the diffusant flows out (for volume diffusion for outflow $F = s\delta/d$ [65], for the outflow $GB\rightarrow$ (dislocation, sub-boundaries) $F = \delta/(gd)$, for diffusion $TJ\rightarrow GB$ (g = 1) $F_{TL} = \delta/d$). If it is assumed that the distance between the TJs is $\sim d$ (the grain size in NC nickel), then the transition from regime $\mathbf{D'_1}$ to $\mathbf{A'}$ takes place at the movement of $t = d^2/D_b$ and equals 25 h for T = 398 K and ~1 h for T = 448 K. This means that in diffusion through the TJ the sequence of the regimes realized in the experiments is: $\mathbf{D'_1}\rightarrow \mathbf{A'}$.

Thus, for nanocrystalline nickel in the temperature range 398–448 K four regimes of grain boundary diffusion ($\mathbf{C'}$, $\mathbf{B'_2}$, $\mathbf{D'}$ and $\mathbf{A'}$) can occur irrespective of the nature of the defects responsible for diffusion mass transfer in diffusion of copper from the surface: they are GB or TJ. As already mentioned, the lowest coefficient of diffusion through the defects (GB or TJ) D_n, corresponding to the experimental ratio $\bar{C}/\bar{C}_0$ is determined in the regime $\mathbf{C'}$ [65]. Consequently, the calculated values of D_n [168] are only the lower estimate of the actual diffusion coefficient with the temperature range 398–448 K.

The temperature dependence of D_n of copper in NS nickel was used to determine [157] the activation energy of grain boundary diffusion Q^n_b = 43.5 kJ/mol (Fig.3.2). This value is less than half the value for grain boundary diffusion in coarse-grain nickel. As discussed in [64, 155, 156], where investigations were carried out into diffusion in nanocrystalline nickel containing pores at the grain boundaries, the activation energy of grain boundary diffusion in NS nickel is similar to the activation energy of diffusion through free surfaces, regardless of the absence of pores in nickel [157]. In addition to this, the activation energy of grain boundary diffusion of copper in NS nickel coincides with the activation energy of grain boundary self-diffusion in NC nickel (46 kJ/mol) [156].

It should be mentioned that in calculations of the activation energy [168] another value of the coefficient of grain boundary diffusion of copper in NS nickel, measured at T = 573 K, was taken into account. In this case, migration of the grain boundaries was detected during diffusion annealing. The final distribution of the grain size is bimodal – in some regions the mean size is 0.4–0.8 μm, in others it does not change. In [168] the diffusion coefficient $D_n = 1.4\cdot10^{-11}$ m²/s was estimated for the grain boundaries migrating at a rate of $V_m = 7\cdot10^{-11}$m²/s corresponding to the regions of the material with the maximum grain size. The identical situation with the bimodal size distribution of the grains was described in [161]. According to the assumptions made in [161], it was assumed that the calculations in [168] gave considerably higher (by two or three orders of magnitude) actual coefficients of grain boundary diffusion which determined the measured depth of penetration of the diffusant. As a result, in [168] when taking into account the value D_n (573 K), the resultant value of the activation energy of grain boundary diffusion of copper in nanostructured nickel was obviously too high.

As already mentioned, the grain boundaries in NS metals are characterised by a high non-equilibrium state (high energy state) whereas in NC and coarse-grain metal the state of the grain boundary is close to equilibrium. Therefore, the values of D_n in NS nickel are considerably higher in comparison with those for NC and coarse-grain nickel (Table 3.1). It may be assumed that the grain boundary diffusion of the impurity atoms in the investigated materials is determined mainly by the state of the grain boundaries and to a lesser degree depends on the amount of the grain boundary phase [157]. In transition of the grain boundaries of the NS nickel to the equilibrium state, the values of D_n should decrease. In [157] it was shown that according to the results of transmission electron microscopy, the transition of the grain boundaries of NS nickel to the equilibrium state starts during annealing at 523 K (for more details see Chapter 4). Thus, after annealing at 523 K for 1 h the grain size of NS nickel did not change, and after annealing for 1 hour at 623 K it increased to 1–3 μm.

After diffusion annealing at 423 K for 50 h, no copper is found in nanostructured nickel pre-annealed at 623 K even at a depth of 1–2 μm [157]. In NS nickel, pre-annealed at 523 K for 1 h, after diffusion annealing in these conditions copper penetrates to a depth of 3–4 μm [157]. In both examined cases, because of the small depth of penetration of copper into nickel it is not possible to determine accurately the values of D_n. However, the estimates of D_n of copper for NS nickel, pre-annealed for 1 h at 523 K, using equation (2.19) give the maximum value of $2.8\cdot10^{-18}$ m^2/s [157], which is four orders of magnitude lower than in the initial condition (Fig.3.2). The controlling role of the non-equilibrium state of the grain boundaries in increasing the coefficients of grain boundary diffusion of the investigated nickel is confirmed by the decrease of D_n at T = 423 K by at least two orders of magnitude observed after holding for three years at room temperature [157].

Recently, the authors of the monograph investigated the grain boundary diffusion of copper in nanostructured nickel produced by torsional deformation by torsion (P = 66 Pa, d = 0.2 μm). Layer analysis was carried out by ion etching and the concentration of this solute in the layer was determined by the method of high-intensity Mössbauer spectroscopy.

The results show that the coefficient of grain boundary diffusion of copper at T = 373 K (95 h annealing time) in nanostructured nickel is $\approx 10^{-12}$ m^2/s, which is 1–2 orders of magnitude higher than the appropriate coefficients extrapolated for coarse-grain nickel from high temperatures (Fig.3.2).

The relatively large difference in the diffusion parameters, obtained recently and presented previously in [157], may be related, firstly, to the different methods of formation of the nanostructure (equal-channel angular pressing and torsional deformation). Secondly, the presence in the material of the impurities, both those associated with the composition of the initial material and those introduced in the process of severe plastic deformation, may strongly influence the diffusion parameters.

Thus, the following main results of investigations of the relationships and mechanisms of diffusion in the nanocrystalline and nanostructure materials have been obtained.

1. In nanocrystalline metals, produced by the methods of gas condensation, electrodeposition, etc, examination showed low-temperature anomalies of grain boundary diffusion: a large (by several orders of magnitude) increase of the coefficients and a decrease (almost halving) of the activation energy of diffusion in comparison with those for the coarse-grained condition. These special features are thermally unstable and are eliminated in recovery processes (migration of grain boundaries).

2. Physical models describing the unusual properties of the grain boundaries in the nanocrystalline materials have been proposed: a) the model of the 'gel-like' condition of the boundaries, based on the results of computer modelling (the correlation effect in the ultrafine grain condition, formed as a result of the small grain size and the absence of relaxation effects are taken into account); b) the cluster model based on direct electron microscopy examination of the structure in which a special role is played by the boundaries of complexes of adjacent grains 'clusters' which contain grain boundary pores and the concentration of the main part of the excess (free) volume of the nanocrystals. These models explain the low-temperature anomalies or grain boundary diffusion described in paragraph 1.

3. In the nanostructured metals, produced by severe plastic deformation (like in the nanocrystalline metals), similar thermally unstable low-temperature anomalies of grain boundary diffusion take place. Direct diffusion experiments show that the high rates of mass transfer in these metals are associated with the non-equilibrium high energy state of the interfaces and not with the grain size. Since these anomalies and the controlling role of the state of the grain boundaries for the given class of materials are also found at grain sizes of the order of 0.3–0.5 μm (the presence of strong correlation effects is unlikely) and in the absence of grain boundary porosity, the currently available models of grain boundary diffusion mechanisms (paragraph 2) are not suitable for explaining the above mentioned special features of diffusion, at least for nanostructured materials produced by the effect of severe plastic deformation.

TABLE PAGE 99 PAGE 116, 120, 121

CHAPTER 4

THERMAL STABILITY OF THE STRUCTURE AND MECHANICAL PROPERTIES OF NANOSTRUCTURED MATERIALS

As already mentioned, the effect of severe plastic deformation results in the formation of ultrafine-grained nonequilibrium structures in the metals and alloys. These structures are characterised by the high density of lattice and grain-boundary dislocations which generate the fields of long-range elastic stresses. Consequently, there are large atomic displacements from the nodes of the ideal crystal lattice and, consequently, the nanostructures, produced by the methods of severe plastic deformation, are characterised by high stored energy and are metastable. It is therefore important to examine the relationships of the evolution of these structures at elevated temperatures and also in the conditions of the combined effect of temperature and loading.

In this chapter, the results are presented of the experimental investigations of the effect of heating on the evolution of nanostructures, produced by severe plastic deformation.

4.1. THERMAL STABILITY OF THE STRUCTURE

It is well known that the growth of grains in the nanostructured materials, produced by severe plastic deformation or other methods, starts at temperatures close to $0.4T_m$ and even lower [7, 172, 173]. Examination of the nature of low thermal stability is of considerable importance for the development of methods of increasing this stability. On the other hand, examination of the evolution of the structure during annealing makes it possible to detect the nature of high internal elastic stresses and their relationship with the lattice defects and nonequilibrium condition of the grain boundaries. Of special interest is the ordering and formation of supersaturated solid solutions detected in many alloys [174, 175]. The thermally activated processes of evolution of the microstructure in the nanomaterials, produced by severe plastic deformation, has been the subject of examination in a number of recent studies [174, 176–185]. Examination of the structure using the method of transmission electron microscopy and x-ray diffraction analysis has been carried out on pure metals: Cu [193, 184], Co, [180], Ni [179, 182], Fe [175], Ti [196], industrial steels [176], intermetallics [174, 177] and composites [187]. The results of these investigations were used to determine the temperature ranges characterised by the occurrence of the processes of recovery and recrystallisation. Nevertheless, the relationships governing these processes have been studied insufficiently. It is therefore interesting to examine the evolution of certain properties during annealing [8, 178, 184], in particular, electrical resistance whose value depend strongly on the degree of defectiveness of the structure.

We examine the main special features, detected in the examination of the thermal stability of various classes of nanostructured materials, produced by severe plastic deformation.

Pure metals

The structure of pure nickel, subjected to severe plastic torsional deformation (five revolutions at room temperature, P = 7 GPa) [178], is characterised by very fine equiaxed grains with the mean size of approximately 100 nm, containing a high density of lattice dislocations (Fig. 4.1). The complicated diffraction contrast indicates the presence of internal elastic stresses. The grains are characterised mainly by high-angle boundaries, confirmed by the appearance of diffraction patterns, containing a large number of reflections, distributed around the circles. These data are in agreement with the results of other examinations of the structure of nickel after severe plastic torsional deformation [37, 188].

Detailed experiments [188] also show that the examined structural condition corresponds to high values of electrical resistance (Fig. 4.2 *a*, $\rho_{4.2}$ = 0.26 μOhm·cm) and elastic strains (see Fig. 4.2 *b*, $\langle\varepsilon^2\rangle^{1/2} \cong 0.076\%$). This confirms the high-density of defects and high internal stresses. This condition is also

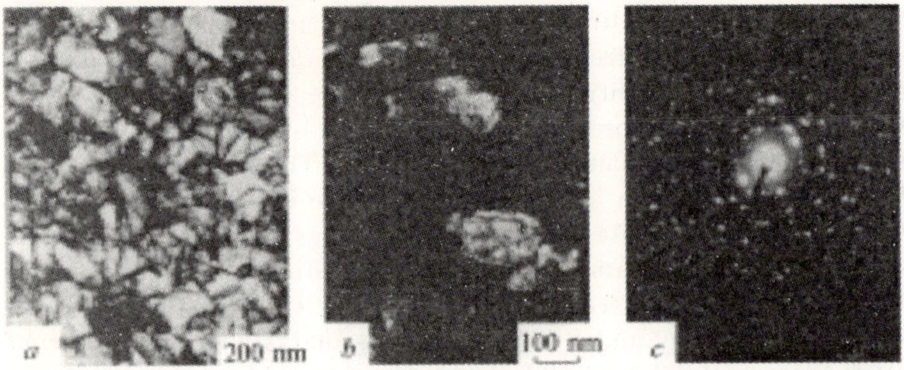

Fig. 4.1. Bright field (*a*) and dark field (*b*) images of the nanocrystalline structure of Ni after severe plastic deformation and the diffraction pattern from the area of the structure with the size of 0.5 μm^2 (*c*).

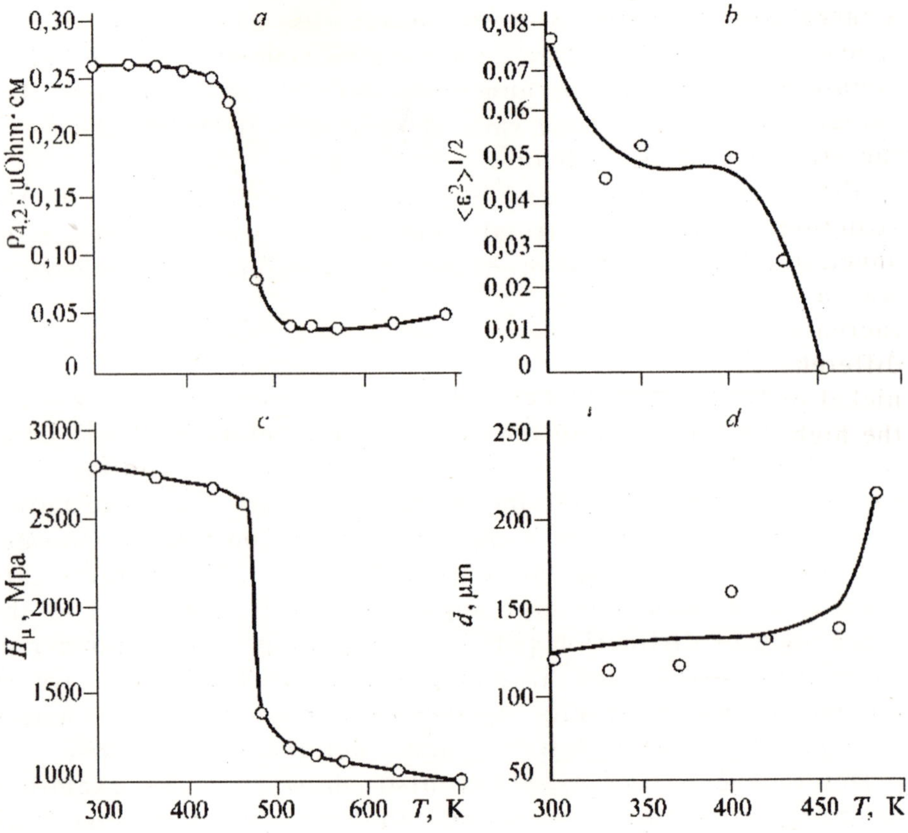

Fig. 4.2 Dependence of electrical resistance at 4.2 K $\rho_{4.2}$ (*a*), internal elastic strain $<\varepsilon^2>^{1/2}$ (*b*), microhardness H_μ (*c*) and the grain size *d* (*d*) on the temperature of annealing for 30 min in Ni subjected to severe plastic deformation in torsion.

characterised by maximum microhardness H_μ = 2850 MPa.

During heat treatment of nickel, subjected to severe plastic deformation, examination showed a decrease of the coercive force, determined by large changes in the structure (Fig. 4.3) [179]. The initial stage of heating is characterised mainly by the disappearance of dislocations inside the grains, i.e. recovery takes place. This process already started at room temperature. For example, after holding for 1 month, the density of dislocations decreased from 10^{15} to 10^{13} m^{-2}. A similar result was also obtained in the case of holding at T = 373 K for 30 min. Heating to a temperature of 473 K results in the complete disappearance of dislocations during 1 min. At T = 473 K banded contrast appeared at the boundaries of some grains, indicating the recovery of the structure of the grain boundaries (Fig. 4.4). The process of recovery of the structure of the boundaries was quite extensive at a temperature of 573 K in a large part of the volume of the specimen. Examination also revealed the formation of a recrystallised structure with the simultaneous growth of grains and the appearance of a certain amount of annealing twins. At T = 673 K and higher temperatures, the structure of the steel became completely recrystallised and further growth of the grains took place. Similar evolution of the structure of nickel, subjected to severe plastic deformation, in annealing was also reported in [188]. However, in this case there was a tendency for the recovery of internal stresses and the increase of the grain size at lower temperatures. The observed differences may be caused by the 'purer' chemical composition of nickel or by the higher stored energy of deformation as a result of the higher pressure, applied in severe plastic torsional deformation

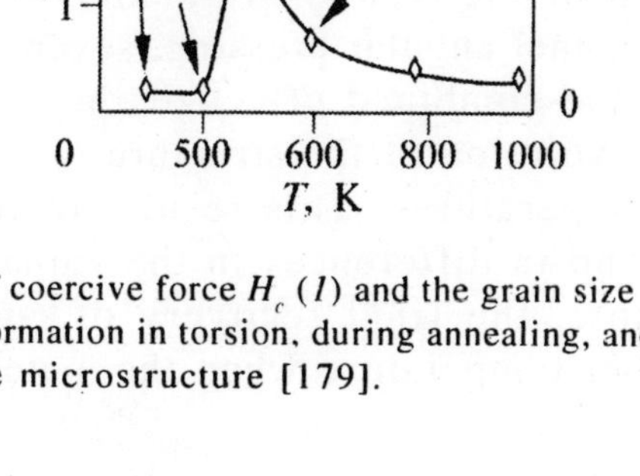

Fig. 4.3. Variation of coercive force H_c (*1*) and the grain size d (*2*) in nickel, subjected to severe plastic deformation in torsion, during annealing, and the diagram illustrating the evolution of the microstructure [179].

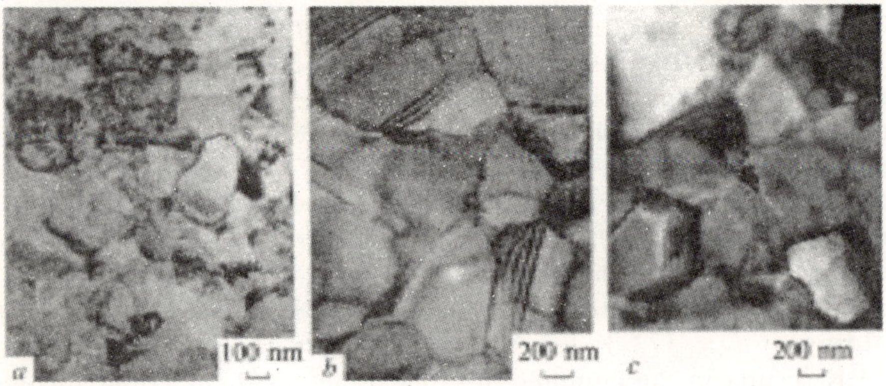

Fig. 4.4. Bright field image of the structure of specimens of nickel in the initial deformed condition (*a*) and annealed for 0.5 h at T = 473(*b*) and 673 K (*c*).

Examination of the properties of nickel subjected to severe plastic deformation in the process of evolution of the structure in annealing showed that there are three main ranges of the variation of electrical resistance and microhardness [188]. In the initial stages of recovery of the structure, accompanied by a decrease of dislocation density and internal stresses, there is a small change of electrical resistance and microhardness (Fig. 4.2*a, b*). During recrystallisation, these parameters greatly change. The subsequent process of grain growth has only a slight effect on these characteristics. At the same time, the magnetic parameters change in a different manner. It has been found [189] that magnetic susceptibility slightly increases with increasing annealing temperature. At temperatures above 470 K the rate of increase increases. This corresponds to the occurrence of recrystallisation. Coercive force H_c [179] is sensitive to the process of evolution of the microstructure up to the grain size of several microns (Fig. 4.3). Above this size, the coercive force becomes less sensitive to grain growth. Remanent magnetisation changes in a complicated, non-monotonic manner. This was discussed in detail in [179].

The method of production has a strong effect on the evolution of the nanostructure in annealing and on the properties of the material. Comparison of the results of examination of Ni and Ti, produced by equal-channel angular pressing, severe plastic torsional deformation and by the combined effect of these processes [190–194] shows that the evolution of the structure in these alloys takes place at different temperatures. This results in the formation of different structures and in differences in the value of the coercive force H_c. For example, the final recovery of coercive force H_c takes place at a higher temperature when the material is subjected

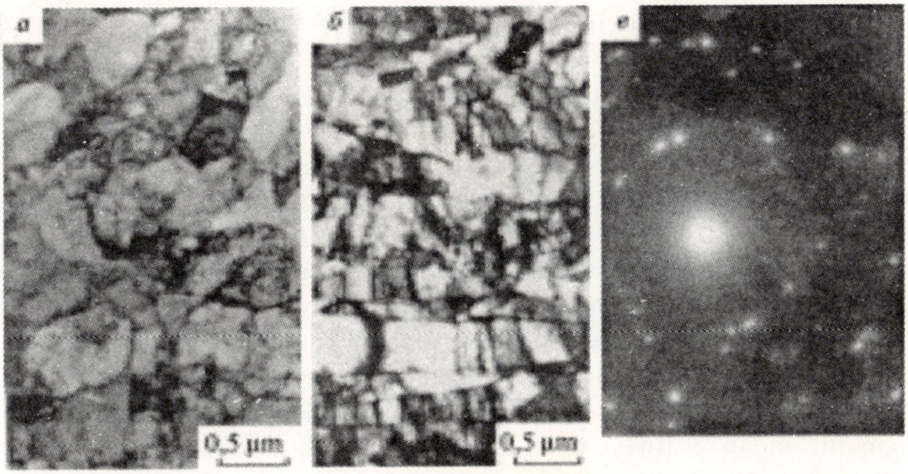

Fig. 4.5. Bright field image of the structure (*a, b*) and the microdiffraction pattern (*c*) of the specimens of nanostructured nickel, produced by the effect of severe plastic deformation by the method of equal-Channel angular pressing.

to equal-channel angular pressing and then to torsion.

The authors of [186, 191] found that the pure metals may be characterised by the formation, after equal-channel angular pressing, of nanostructures with the grains elongated along the direction of pressing. Typical bright field images of such a microstructure for nickel showed the equiaxed grain structure with the grain size of ~300 nm only in the transverse cross-section in relation to the direction of pressing (Fig. 4.5*a*) [186]. In the longitudinal direction in relation to the pressing direction, the grains were elongated with the coefficient of non-equiaxiality of the order of 2 and more and the grain size of 300–600 nm (Fig. 4.5*b*). The volume of the grain is often characterised by extinction contours indicating the presence of internal stresses which, as in the case of the equiaxed nanostructures, may be associated in particular with the defective structure of the grain boundaries. The structure contains grains with the boundaries with the banded contrast, typical of the equilibrium condition of the grain boundaries, but the fraction of these boundaries does not exceed 2%. The electron diffraction patterns contain a large number of inflections distributed around the circumference, with some of them characterised by considerable azimuthal blurring (Fig. 4.5*c*). This form of the electron diffraction patterns indicates the presence of high-angle misorientation between the elements of the nanostructure and the presence of elastic stresses in the individual grains. The direct measurements of the misorientation of the individual boundaries of the structure with elongated grains, taken by the authors of [186] show that the elongated grains are usually restricted by the high-angle boundaries. The volume of the majority of the grains contains low-angle boundaries of the subgrains. These boundaries are characterised by a network or network-cellular dislocation structure. In the grains with the size close to the mean size (d ~300 nm) the low-angle boundaries have the misorientation angle of approximately 1°. These boundaries are almost perpendicular to the plane of distribution of the high-angle boundaries. In the larger grains, the low-angle boundaries are more elongated and the misorientation angle of these boundaries is approximately 10°. In the grains with the size smaller than the mean grain size of the low-angle boundaries are usually not found. Evidently, these special features maybe regarded as common for the structures, formed under the effect of severe plastic deformation by the method of equal-channel angular pressing.

Examination of the effect of annealing on the structure of nanostructured nickel, produced by equal-channel angular pressing, carried out in [191] shows that annealing in the temperature range 393–548 K does not lead to any significant change of the structure, including the grain size, determined by electron microscopy examination. However, after annealing at T = 523 and 548 K the number of the grain boundaries with the banded contrast increases to 10–20% and the extinction contours inside the grains, surrounded by these boundaries, disappear. This indicates the development of the processes of stress relaxation in these grain boundaries at a constant grain size. The decrease of the internal stresses may be determined by the redistribution and annihilation of the grain boundaries and lattice dislocations at the grain boundaries and in the boundary regions [174]. It should be mentioned that the incubation period of the process of grain growth was previously detected in, for example, nanostructured molybdenum [195] and iron [175], produced by severe plastic deformation. After annealing at a temperature of 573 K the grain size in the individual areas of the specimen increased to 0.4–0.8 μm. Individual dislocations were found inside such grains (Fig. 4.6*a*). The electron diffraction patterns taken from these areas have the form typical of coarse-grained nickel (Fig. 4.6*b*). At annealing temperatures higher than 573 K recrystallisation took place in the specimens. The size distribution of the grains in such a structure is bimodal (Fig. 4.6*c*, regions of the type 1 and 2). For example, after annealing at T =

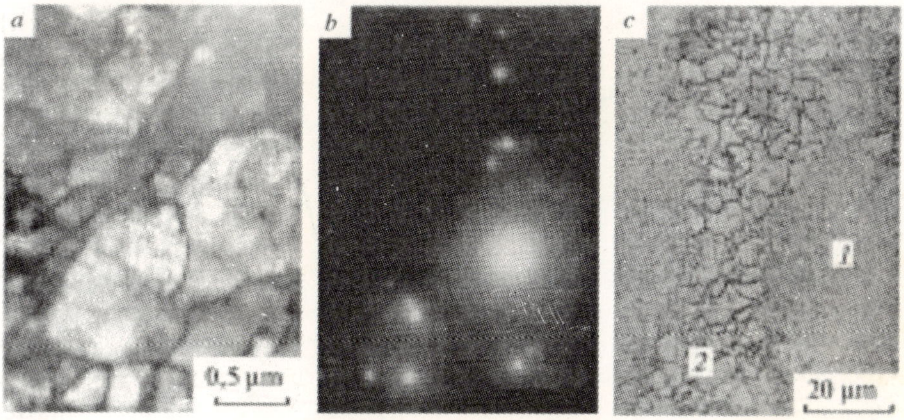

Fig. 4.6. Structure (*a, c*) and the microdiffraction pattern (*b*) of the specimens of nanostructured nickel after annealing at 573 (*a, b*) and 623(*c*) K. 1, 2) see the text.

623 K the mean grain size in the type 1 region (the volume fraction 70%) is 0.8–0.9 μm, and in the regions of type 2 (volume fraction 30%) it is 3.5 μm. With increasing annealing temperature the mean grain size in the regions of type 2 increases, and in the regions of type 1 it is almost constant, but the volume fraction of the regions of type 1 decreases. After annealing at T = 673 K the structure of nickel becomes completely homogeneous and the size distribution of the grains in the structure is close to a normal logarithmic distribution. The dependence of the mean grain size of nanostructured nickel on annealing temperature as shown in Fig. 4.7 (curve 1 – region of type 1; curve 2 – region of type 2).

The measurement of microhardness (H_μ) of nanostructured nickel shows that microhardness changes in a non-monotonic manner with annealing temperature (Fig. 4.8). After annealing in the temperature range up to 523 K, the values of H_μ are uniform in the specimen and decrease from 2.68 to 2.20 GPa (mainly after annealing at T = 523 K). Annealing at a temperature of 573 K results in the formation of heterogeneity of H_μ in the specimen. In the regions with the recrystallised structure H_μ rapidly decreases to 145 GPa, whereas in the regions in which recrystallisation is not detected in optical examination, H_μ remains at 2.20 GPa. With a further increase of annealing temperature to 773 K the value of H_μ gradually decreases to 1.2 GPa and becomes uniform throughout the specimen.

The variation of specific electrical resistance with annealing temperature is also non-monotonic (Fig. 4.7, 3). Figure 4.7 shows

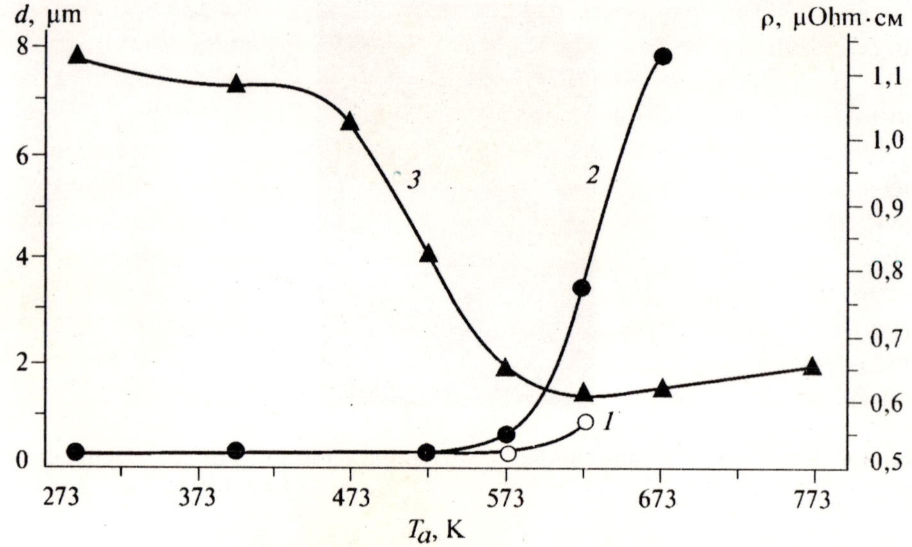

Fig. 4.7. The dependence of the mean the grain size (*1*, *2*) and the specific resistance (*3*) at T = 77 K of nanostructured nickel on preliminary annealing temperature.

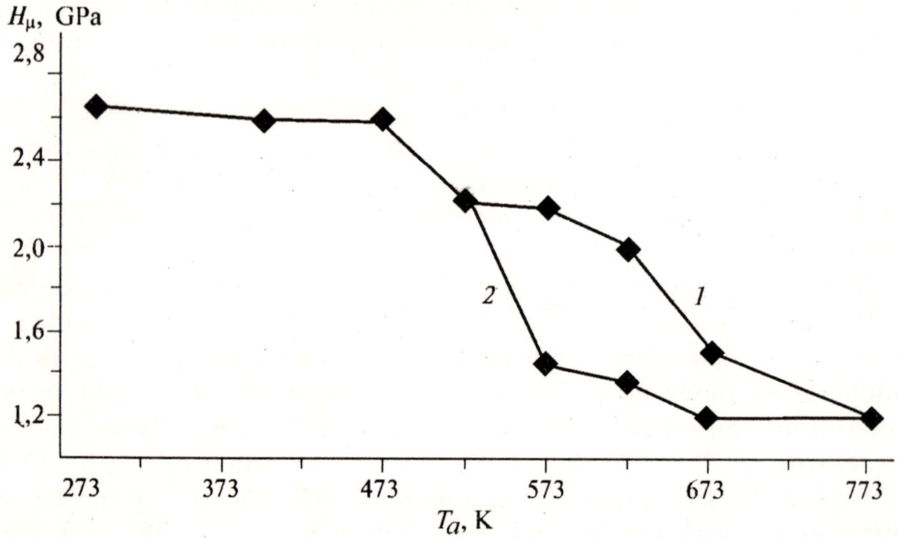

Fig. 4.8. Dependence of the microhardness of nanostructured nickel on preliminary annealing temperature. *1*) type *1* region; *2*) region *2*.

that after annealing in the temperature range 273–473 K the value of ρ is almost constant. A large decrease of ρ to the appropriate value of coarse-grained nickel is recorded only after annealing at a temperature of 523–623 K. The temperature of the start of the rapid decrease of this parameter is in the temperature range where there is no grain growth and, as already mentioned, only the processes of recovery in the structure of the grains boundary should take place (Fig. 4.7, 3).

Thus, the results of examination of the temperature evolution of the structure and properties of nanostructured nickel, produced by severe plastic deformation, show that heating of this material is accompanied by complicated structural changes associated with the development of the processes of recovery, recrystallisation and grain growth. Evidently, the nature of recovery is associated mainly with the redistribution and annihilation of dislocations at the boundaries and in the body of the grains leading to a decrease of the internal stresses (Fig. 4.2*b*).

On the basis of the data one can propose the following sequence of the evolution of the nanostructure in heating and nickel subjected to severe plastic deformation (Fig. 4.3). In the material, the grains size remains constant up to a temperature of 473 K but the recovery of the structure, associated with the redistribution and annihilation of the dislocations, takes place inside the grains and at of the grain boundaries. Recrystallisation and rapid grain growth start at high temperatures and they may be of anomalous nature if the structure contains a residual density of dislocations. However, two conclusions should be made regarding the given scheme of evolution of the nanostructure.

1. The sequence of the processes, taking place with increasing temperature, is of course not absolutely strict. The temperature ranges in which these processes take place slightly overlap and this depends on the chemical purity of the metal, regimes and conditions of severe plastic deformation.

2. The process referred to as 'recrystallisation' in fact differs from the process usually detected in the annealing of cold-deformed metals. 'Conventional recrystallisation' is the nucleation and growth of new perfect grains as a result of the deformed matrix in which recovery takes place. During conventional recrystallisation, the structure is bimodal because of the co-existence of large perfect grains and fine imperfect cells. The process of recrystallisation may lead to changes of both the grain size and grain orientation (texture).

In contrast to the previously described case, in the nickel subjected to severe plastic torsional deformation, the recovery of the structure results in the formation of grains with a single-modal distribution in respect of the size and orientation, similar to the orientation existing in the deformed specimen, i.e. the texture does not change. This process is similar to the process referred to as recrystallisation *in situ* encountered in certain steels [196]. If for some reason (for example, as a result of precipitation) the migration of new grain boundaries is difficult, only recovery can take place and the resultant structure contains mainly small but misoriented grains. In the process of recrystallisation *in situ* the deformation texture does not change to any large extent. The texture changes only in the process of subsequent grain growth.

The elongated structure, identical with that in nanostructured nickel, forms in equal-channel angular pressing and in titanium VT1-0 (composition in wt%: 0.12 O, 0.15 Fe, 0.07 C, 0.04 N, 0.01 H) (Fig. 4.9). The results of electron microscopic examination show [196] that after equal-channel angular pressing this material is characterised by the bimodal size distribution of the grains. The mean size of the small grains is 0.3–0.35 μm. The size of the larger grains varies up to 1 μm and the fraction of these grains, according to approximate estimates, may reach 40%. Extinction contours, indicating the presence of elastic stresses, form inside the gains. During heating, large changes in the microstructure are detected only after annealing at a temperature of 573 K and higher. This relates mainly to the decrease of elastic stresses, as indicated by the formation of a banded contrast at the grain boundaries, characteristic of the equilibrium condition of the grain boundaries, and the disappearance of the extinction contours inside the grains with these boundaries. The growth of grains in nanostructured

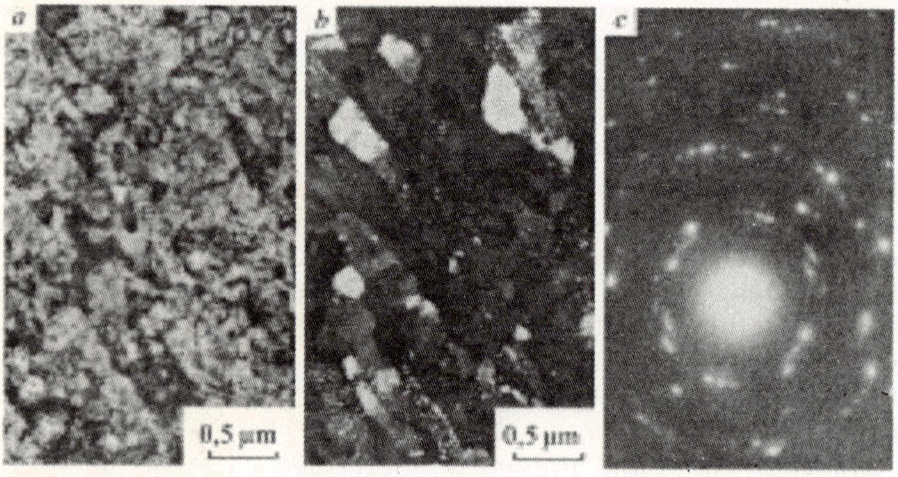

Fig. 4.9. Bright field (*a*) and dark field (*b*) images of the structure and the microdiffraction pattern (*c*) of the specimens of nanostructured titanium, produced by severe plastic deformation using the equal-channel angular pressing method.

titanium starts after annealing at a temperature of 723 K for in individual sections, the grains size increases to 1.5 μm together with the existing non-recrystallised areas. The degree of elongation of the structure in the longitudinal section greatly decrease. Complete recrystallisation is observed after annealing for 1 h at a temperature of 773 K.

A special feature of the electron diffraction patterns of the specimens of VT1-0 titanium in the nanostructured condition is the frequent presence of extra-reflections in the positions [0001], forbidden for titanium by the structural factor. The grains show the coherent disc-shaped precipitates resulting in a deformation contrast as a result of elastic distortion of the lattice of the matrix (Fig. 4.10*a*). The diameter of these precipitates is approximately 10 nm. The dark field micrographs in the 'light' of the extra-reflections confirm that the [0001] extra-reflections form from the precipitates (Fig. 4.10*b*).

The examined titanium VT1-0 of commercial purity contains a relatively large amount of interstitial impurities – nitrogen, oxygen and carbon. The total amount of these impurities is almost 0.9 at%. Each of these elements forms interstitial phases with titanium and, consequently, it may be assumed that the process of production of nanostructured titanium is accompanied by the formation of precipitates of these phases. However, the detected extra-reflections do not correspond to any of the known interstitial phases based on titanium.

Carbon is the interstitial impurity with the lowest solubility in titanium at temperatures below 773 K [197]. In addition to this, the diffusion coefficient of is higher than that of oxygen and nitrogen. The condition of carbon at temperatures below 773 K has not been determined reliably. Since the carbon content of the investigated

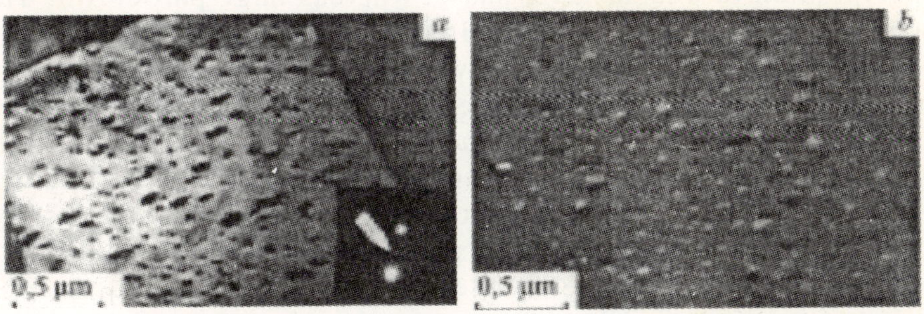

Fig. 4.10. Microstructure of nanostructured titanium after annealing at T = 773 K for 30 min. *a*) bright field micrographs, extra-reflection [0001] is indicated by the arrow; *b*) the dark field micrographs in the given reflection.

material is 0.3 at% (higher than the limit of solubility), it has been assumed that the precipitates represent precipitates of a metastable phase based on carbon.

The appearance of extra-reflections of type (0001), forbidden by the structural factor for α-Ti, is explained in [198] on the basis of the model of distribution of carbon in the lattice of coherent precipitates. Carbon occupies octahedral interstitials [197] in the inter-nodal plane, parallel to the basal plane of the HCP lattice of α-Ti. It may be expected that the next interstitial plane is free of the carbon atoms. Increase of the carbon concentration of titanium increases the lattice parameters of α-Ti, and the value of the parameter *c* increases more appreciably than the value of *a* [197]. The high concentration of carbon in comparison with the surrounding solid solution leads to the formation of coherent elastic stresses which cause the formation of a deformation contrast. The solubility of carbon in the α-phase increases with the increase of temperature to 1073–1123 K and, consequently, the specimen of VT1-0 titanium was quenched in vacuum from temperatures of 1073–1103 K. Quenching was carried out by throwing a titanium sheet with a thickness of 0.5 mm on a cold metallic plate after heating in a molybdenum heater in VUP-5 vacuum station. After quenching, the structure of titanium contained only plate-shaped precipitates of titanium carbide TiC, as indicated by the analysis of electron diffraction patterns. The specimen was aged for 1 h at a temperature of 773 K. The aged specimen contained disperse precipitates of a coherent phase with the deformation contrast and the particles of TiC which resulted, on the microelectron diffraction patterns, in reflections corresponding to the TiC lattice. In examination of the dark field in the reflections of TiC there was no glow of disperse precipitates, i.e. ageing was accompanied by the precipitation of a metastable phase, whereas large particles of TiC were retained from the quenched condition.

Thus, the phase formed at low temperatures is metastable and at high temperatures is replaced by stable precipitates of TiC carbide. Carbon in the precipitates is in octahedral interstitials situated between two adjacent planes of the basis of the HCP lattice of titanium resulting in the formation of extra-reflections of the type 0001 on the electron diffraction pattern.

The structure of copper after severe torsional plastic deformation differs from that of nickel. The mean grain size of copper prepared in this manner [37, 173, 183] is usually slightly larger (150 nm). In addition to this, the specimens of superpure copper contained a certain amount of recrystallised larger grains (0.5 μm). The grain boundaries are also characterised by mainly high-angle mioriemations.The structure is characterised by a complicated electron microscope contrast indicating extensive distortion of the crystal lattice. The dislocation density inside the grains was 5–$10 \cdot 10^{14}$ cm^{-2}.

Evidently, the differences in comparison with Ni are caused by differences in homologous temperatures (T/T_m) at which deformation took place and the specimen was held at the same temperature after deformation. The temperature of 293 K corresponds to homologous temperatures of 0.17 for nickel and 0.22 in the case of copper. Consequently, the mobility of atoms in the latter case should be higher and this may result possibly in changes in the structure and in differences in the grain size. The increased mobility of the atoms could have caused dynamic recrystallisation.

In copper subjected to equal-channel angular pressing, the mean grain size was slightly larger, 0.2 μm [199]. A large fraction of the grains did not contain any dislocations. In the majority of grains, the dislocations were distributed randomly or sub-boundaries formed. A small part of the grain boundaries was characterised by the contrast typical of the equilibrium grain boundaries.

It has been found that the growth of gains in pure Cu (99.98%), subjected to severe torsional plastic deformation, starts at T = 373 K and is completed at 473 K [200]. In commercial purity copper deformed by a similar method, annealing for 30 min at T = 423 K did not cause any visible changes in the microstructure (grain size, dislocation density, diffraction contrast at the grain boundaries) [201]. The grains started to grow at a temperature of 423 K. Nevertheless, the internal stresses and yield limit greatly decreased in the temperature range 403–423 K (Figs. 4.11 and 4.12). The authors assumed that the internal stresses and, consequently, mechanical strength may be associated with the presence of disclinations at triple points at the grain boundaries. It may be assumed that after disappearance of internal stresses, the flow stress of the specimens is closed to that of copper with a similar grain size produced by a completely different method, namely by the method of gas condensation in an inert medium [201].

The results of a number of investigations [173, 183, 201, 202] show that in low-temperature annealing the processes of recovery in nanostructured copper, produced by severe plastic torsional deformation, start considerably earlier in comparison with coarse-crystalline copper. The temperature in heating is slightly higher than

T = 373 K (Figs. 4.11 and 4.12). The copper, subjected to severe plastic torsional deformation, is also characterised by the use temperature of the start of grain growth, i.e. 423 K

In the copper subjected to equal-channel angular pressing, the relationships of the evolution of the structure in annealing were investigated in [183, 204, 205]. This was carried out by the parallel application of the methods of measuring the speed of propagation of ultrasound waves and internal friction. The results were stored in the already described scheme. Annealing resulted in the formation of grains which did not contain extinction contours and were separated by the grain boundaries which, as assumed, became equilibrium. The recrystallisation temperature (448 K) was, as expected, higher in lower purity copper (99.98%) in comparison with higher purity copper (99.997%) where it was equal to 398 K. These changes in the structure were reflected in the measured properties. The speed of propagation of the ultrasound waves and, consequently, elastic moduli greatly increased. The recrystallised grains continued to grow at higher temperatures. No significant

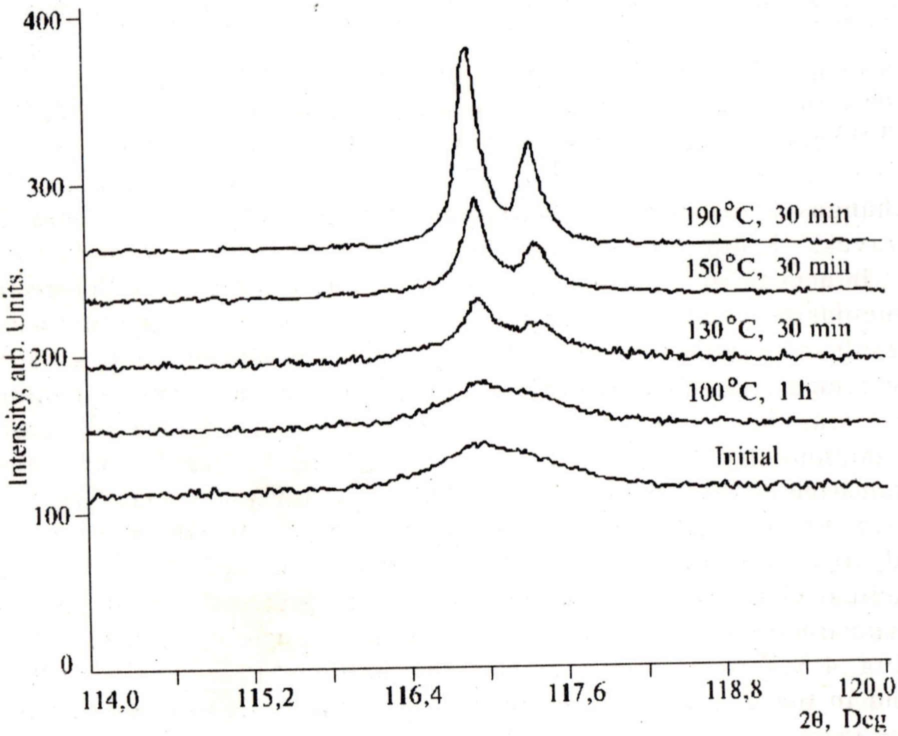

Fig. 4.11. Effect of annealing temperature on the type of x-ray peak (400) of copper subjected to severe plastic deformation by torsion.

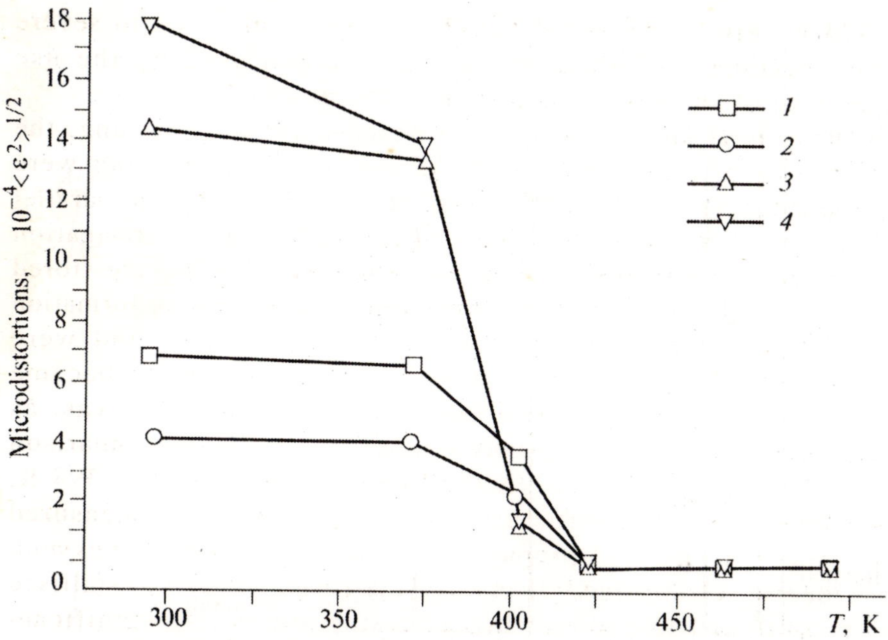

Fig. 4.12. Dependence of the microdistortion of the crystal lattice of copper, produced by severe plastic torsional deformation, on annealing temperature. 1 and 2 were obtained using x-ray peaks (111) and (222), 3 and 4 using the peaks (200) and (400). The curves 1 and 3 were plotted using x-ray spectrum analysis [202], the curves 2 and 4 by the method described in [203].

changes were found in the speed of propagation of ultrasound waves.

In annealing of nanostructured copper, produced by equal-channel angular pressing, at T = 323, 373 and 423 K for 30 min did not result in any large changes on the diffraction patterns (Fig. 4.13*a*) in comparison with the initial condition after equal-channel angular pressing [204]. The maximum intensities of the x-ray peaks remained unchanged. Analysis shows that the grain size inthe investigated directions <101> and <200> did not change but the level of microdistortion of the crystal lattice slightly decreased. Identical annealing for 30 min at T = 473 and 523 K resulted in radical changes on the appropriate x-ray diffraction patterns of nanostructured copper, produced by equal-channel angular pressing (Fig. 4.30*b*). The changes in the maximum and integral intensities, and in the integral broadening of the x-ray peaks were caused by the increase of the grain size and decrease of microdistortions and also by the simultaneous evolution of the crystallographic structure (Fig. 4.14). Comparison of the results with the results of

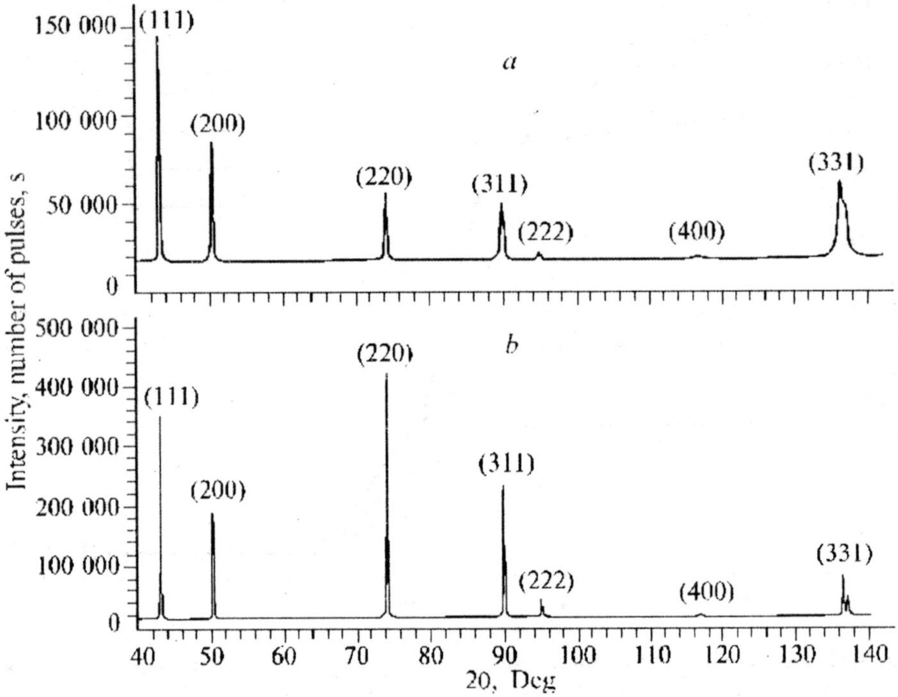

Fig. 4.13. X-ray diffraction patterns of nanostructured copper, produced by equal-channel angular pressing, after annealing at T = 423 (a) and 523 (b) K for 30 min.

measurement of microhardness of the specimens annealed at different temperatures of 30 min, indicate that grain growth started at temperatures above 443 K The states, corresponding to T = 473 and 523 K, are recrystallised. The results published in [204] were used for more detailed analysis of the general appearance of the x-ray diffraction pattern and the evolution of the microstructre in low-temperature annealing at T = 443 K for different periods of time. After annealing for 10, 30 and 60 min, the x-ray diffraction patterns of nanostructured copper, processed by equal-channel angular pressing, correspond to the x-ray diffraction pattern of the initial condition.

After annealing for 1 h 30 min, the broadening of the x-ray peaaks (311), (220), (200) decreased and the appropriate doublets $K_{\alpha 1}$–$K_{\alpha 2}$ started to appear separately (Fig. 4.15). This means that the process of grain growth takes place or is already completed in the appropriate texture components. At the same time, the x-ray peak (331), characterised by maximum integral intensity, changes only slightly comparison with the condition after annealing for 1 h and less. This indicates only the start of the processes of recovery

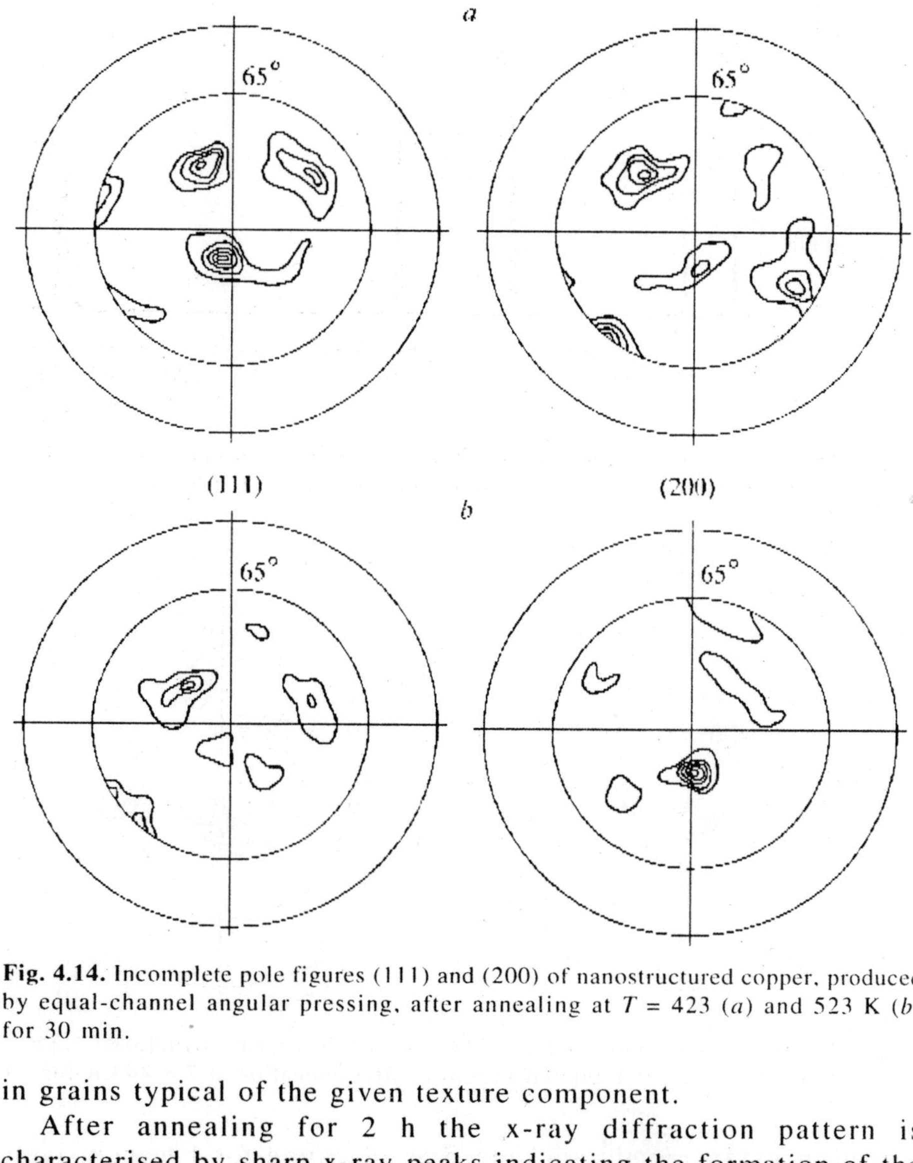

Fig. 4.14. Incomplete pole figures (111) and (200) of nanostructured copper, produced by equal-channel angular pressing, after annealing at T = 423 (*a*) and 523 K (*b*) for 30 min.

in grains typical of the given texture component.

After annealing for 2 h the x-ray diffraction pattern is characterised by sharp x-ray peaks indicating the formation of the coarse-crystalline condition with a low level of microdistortions of the crystal lattice. Analysis [204] of the evolution of the grain size and microdistortions of the crystal lattice during annealing at a temperature of 443 K for different times shows that the increase of the grain size in the direction <111> takes place in stages and rapidly increases after annealing for 1 h 30 min. The large increase of the grain size in the direction <200> is already detected after 1 h and rapidly becomes greater at longer annealing times. On the other hand, the microdistortions of the crystal lattice in the

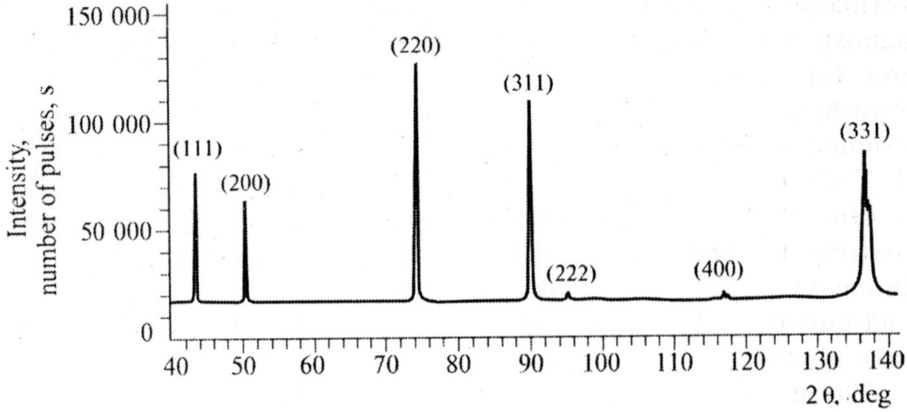

Fig. 4.15. X-ray diffraction pattern of nanostructured copper, produced by equal-channel angular pressing, after annealing at T = 443 K for 1 h 30 min.

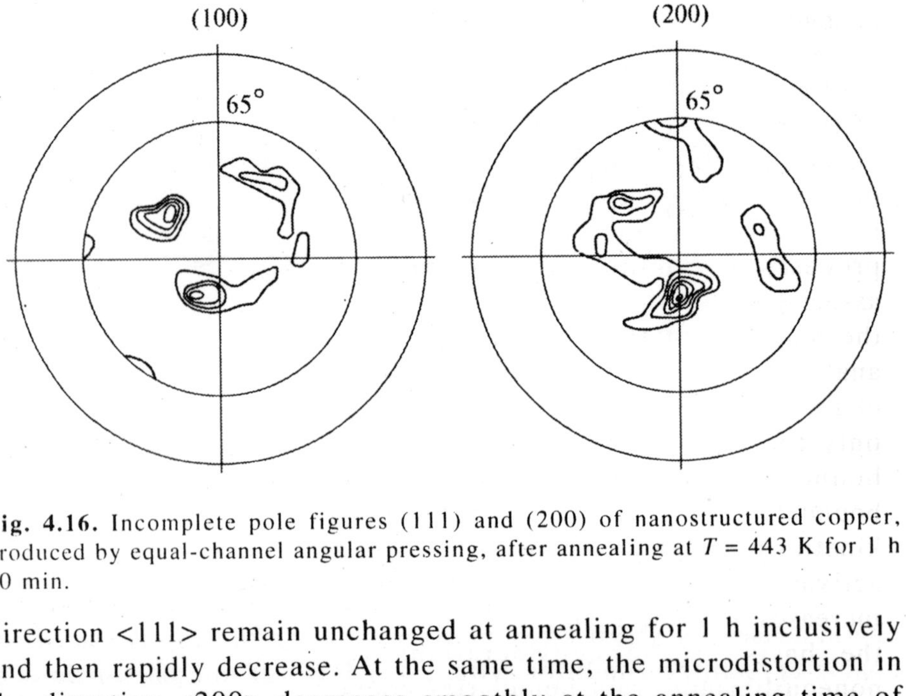

Fig. 4.16. Incomplete pole figures (111) and (200) of nanostructured copper, produced by equal-channel angular pressing, after annealing at T = 443 K for 1 h 30 min.

direction <111> remain unchanged at annealing for 1 h inclusively and then rapidly decrease. At the same time, the microdistortion in the direction <200> decreases smoothly at the annealing time of 1 h and then decreases rapidly. Annealing at T = 443 K for 2 h results in a qualitative change of the type of incomplete pole figures (111) and (200) (Fig. 4.16) in comparison with pole figures corresponding to 10, 30 and 60 min of annealing. The condition, corresponding to the annealing time of 1 h and 30 min, is characterised by the combination of texture maxima, characteristic of the annealing times of 1 and 2 h. These results indicate the formation of grains with a new orientation in the investigated nanostructured Cu, produced by equal-channel angular pressing, in low temperature annealing at $T = 443$ K for 1 h and 30 min. Another confirmation of the formation of grains with a new orientation is the large decrease of microhardness after annealing at $T = 443$ K for 1 h and 30 min.

Nonuniform grain growth, the decrease of the microdistortions, the nonuniform and rapid changes in the integral intensity of the x-ray peaks confirmed the fact that the recovery and growth of the grains, belonging to different texture components, take place nonuniformly.

The experimental results show [173] that the variation of the mean grain size in nanostructured copper, produced by severe plastic torsional deformation and having the grain size of 160 nm, in relation to the annealing time is on the whole similar to the time dependence of grain growth in conventional microcrystalline copper. Nevertheless, grain growth starts at a relatively low temperature, 0.32 T_m. It is well-known that a large decrease of the temperature of the start of grain growth was recorded in many nanocrystalline materials [173]. However, the principle of this phenomenon has not been completely explained. Many investigators have assumed that it is caused by the very high driving force of the grain growth process, determined by the small grain size. Other reasons, associated with the formation of nonequilibrium grain boundaries in the nanostructured materials, have been examined in [173]. The analysis, carried out in [173], shows that the reason for the start of grain growth in nanocrystals at relatively low temperatures is not only the small grain size but also higher mobility of the grain boundaries, determined by the low activation energy of grain boundary diffusion. The comparison of the calculated curves of the kinetics of grain growth with the experimental data shows that the activation energy of grain growth in the nanostructured materials increases with increasing grain size. This fact may be explained on the basis of the considerations regarding the existence of nonequilibrium grain boundaries containing a high density of the introduced grain boundary dislocations. It is well known that the nonequilibrium state may ensure higher mobility of the atoms in the grain boundaries and greatly reduce the activation energy of grain growth [206].

The estimates obtained on the basis of the approximation in [173] for copper give the values of the activation energy of grain growth between 92 and 107 kJ/mole for the grain size of 200 nm and between 82 and 87 kJ/mole for the grain size of 40 nm. During grain growth recovery is detected in the grain boundaries with the latter transferred to a more equilibrium condition. Consequently, the activation energy increases to 107 kJ/mole [207] which is the value characteristic of coarse-crystalline copper.

The evolution of the structure and grain growth in annealing were also investigated in other nanostructured materials produced by severe plastic deformation: Fe [208], Co [180]. After severe plastic torsional deformation with the logarithmic degree of deformation of 6.2, iron contained grains with a diameter of approximately 100 nm containing many bending extinction contours, indicating the presence of high internal stresses. The grain boundaries were also in the non-equilibrium condition. After annealing for 1 h at a temperature of 573 K, the grain size was the same as previously but the grains did not contain internal stresses and dislocations. At T = 773 K examination showed grain growth and the mean size of the grains was 0.3 μm. The grain boundaries were usually characterised by the conventional banded contrast, typical of the equilibrium grain structure. Further grain growth was detected at high temperatures. Microhardness slightly decreased during recovery, rapidly decreased at a temperature of approximately 723 K during recrystallisation, and subsequently slightly decreased during grain growth.

Regardless of the differences in the crystallographic structure, the deformation structure and the scheme of evolution of the structure in the case of iron and nickel were similar [178]. Nevertheless, the temperatures of the start of evolution of the structure differed.

In the case of Co at T = 693 K examination showed certain special features [180] associated with the allotropic transformation of the ε-phase with the HCP lattice to the γ-phase with the FCC lattice, typical of cobalt. In Co [180], severe plastic torsional deformation resulted in the formation of a highly dispersed grain structure containing a high dislocation density, micrograins and lattice defects. In this condition, the initial ε-phase with the HCP lattice was retained. After annealing at T = 673 K there was a small decrease of the density of defects, and at 723 K easily distinguished large grains with a banded contrast at the grain boundaries formed. At T = 773 K the grains continue to grow and the size reaches 5 μm. X-ray diffraction analysis shows that already at T = 693 K the structure contains the χ-phase with the FCC lattice. This indicates that the $\varepsilon \rightarrow \gamma$ transformation in severe plastic deformation of cobalt takes place at a considerably lower temperature in comparison with conventional, coarse-grained Co. Annealing at a temperature of 1073 K resulted in the formation of large grains with a size of 10 μm with a large number of stacking faults. The measurements of coercive force H_c show that the high value of the force after severe plastic deformation gradually decreases down to a temperature of 573 K, i.e. the temperature at which the redistribution of dislocations results in a decrease of the internal stresses. A large decrease of the coercive force was recorded at T = 573–773 K in the area characterised by the occurrence of recrystallisation. At temperatures above 773 K the rate of reduction of the coercive force was lower. Evidently, the decrease is associated with grain growth.

Thus, in all investigated metals, subjected to severe plastic deformation, heating was accompanied by a similar evolution of the nanostructure. The typical feature was the development of the recovery processes, associated with the redistribution and annihilation of the dislocations at the boundaries and in the body of the grains. There were also recrystallisation processes, leading to grain growth, but the sequence of these processes is determined by the chemical composition and nature of metal (the stacking fault energy, the type of crystal lattice), and also by the conditions of severe plastic deformation which form the nature of the initial nanostructures. Detailed investigation should be carried out in this area. It is also important to determine the process controlling the evolution of the structure in heating. In [12, 173] it is assumed that these processes are represented by the structural rearrangement at nonequilibrium grain boundaries whose rate controls the recovery of the structure and the start of recrystallisation. However, the explanation of this process requires further investigations.

Alloys

Attention will be given to special features of the evolution of the structure in heating of nanostructured alloys, produced by the methods of severe plastic deformation. The features of the behaviour of the structure, typical of the alloys, will be studied. As already shown, severe plastic deformation of alloys results not only in excessive refining of the microstructure but also in changes of the phase composition, associated with the formation of supersaturated solid solutions, disintegration of the second phases, possible amorphisation.

The structure of MA8 alloy, subjected to severe plastic torsional deformation at room temperature, consists mainly of very small grains with a diameter of 50–100 nm containing dislocations with a density of 10^{12}–10^{13} m^{-2} (Fig. 4.17*a*) [174]. The grains were usually separated by high-angle boundaries, and many grains were characterised by Moire fringes, formed as a result of double diffraction in the grains with the basal plane parallel to the plane of the foil. In addition to these grains, the structure contained a small number (occupying 5% of the investigated surface) of larger diameter grains (0.11–0.15 μm), containing individual dislocations and characterised by relatively equilibrium grain boundaries.

Annealing for 2 h at a temperature of 413 K resulted in the formation of a structure with more perfect grains with a diameter of 0.18 μm (Fig. 4.17*b*). The investigations carried *in situ* in [174] in the column of an electron microscope during isothermal annealing showed that the formation of new grains is associated with the recovery processes and perfect grain boundaries form as a result of local rearrangement in the nonequilibrium grain boundaries. Thus, there is no stage of nucleation.

The structure of the Al–4% Cu–0.5 Zr alloy, subjected to severe plastic deformation [44], consisted of almost dislocation-free randomly disoriented crystals (grains with the size of 0.2 μm) containing bend extinction contours and having the boundaries without a banded contrast. Annealing for 2 h at T = 433 K resulted in the formation of a structure with the same grain size but with

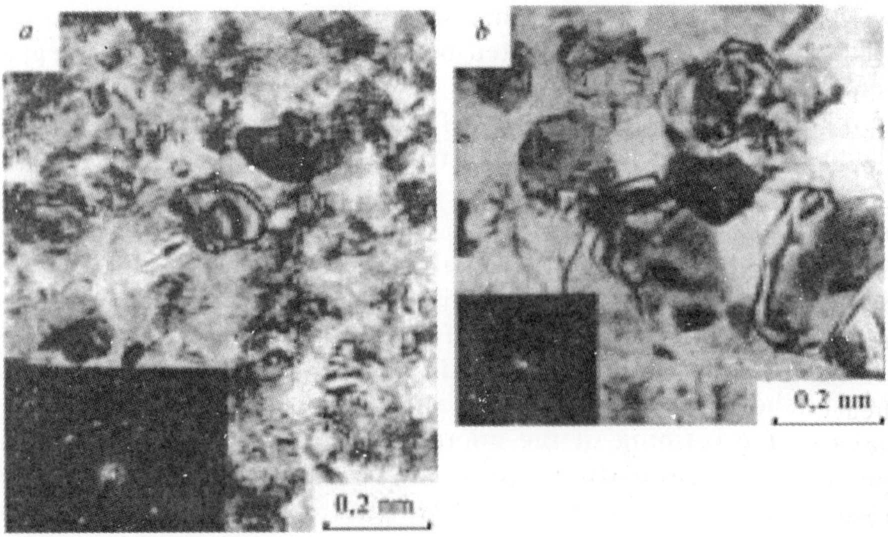

Fig. 4.17. The microstructure of MA8 alloy. *a*) after severe plastic deformation in torsion (true logarithmic strain e = 6) at T = 293 K; *b*) after additional annealing at T = 413 K for 2 h.

more perfect grain boundaries characterised by the conventional banded contrast. Subsequent deformation at the same temperature revealed superplastic behaviour and resulted in an increase of the grain size up to 0.5 μm. Additional results were reported in [209] in which a considerably smaller grain size, approximately 0.1 μm, was detected after severe plastic deformation of the Al–4%Cu–0.5% Zr alloy (Fig. 4.18*a*). The annealing of this material for 1 h at T = 343–353 K resulted in a decrease of the number of extinction contours in the grains. After annealing for 1 h at T = 363 K the structure of the grain boundaries improved, and a third of the grain boundaries showed a banded contrast. At T = 373 K, examination showed the start of grain growth with the mean size of the grains reaching 0.17 μm (Fig. 4.18*b*). Dilatometric investigations [209] in the same temperature range showed a decrease of the length of the specimens. Analysis of the temperature dependence of the variation of the initial length made it possible to calculate the enthalpy, equal to 36 kJ/mole. This is approximately 50% of the value of grain boundary self-diffusion in aluminium. The authors of [209] explained this fact by the transformation of nonequilibrium grain boundaries to equilibrium ones.

In the Al–1.5% Mg solid solution, severe plastic torsional deformation resulted in the formation of a structure very similar to that formed in the previously discussed Al–Cu–Zr alloy, but with a smaller grain size (0.15 μm instead of 0.2 μm) [210]. Annealing up to T = 373 K did not result in any changes in the grain size. At a temperature of 393 K the extinction contours disappeared. These contours existed previously in the body of the crystals. The grain size slightly increased (up to 0.18 μm) and a banded contrast appeared in approximately 50% of the grains. At high temperatures grains started to grow. The microhardness measurements showed that the microhardness was maximum after annealing at T = 373 K. At highed temperatures, microhardness decreased as a result of grain growth in accordance with the Hall–Petch relationship.

The processes, taking place in the supersaturated Fe–O alloy in heating, were investigated in [175]. This alloy was produced by severe plastic torsional deformation of an oxidised submicron iron powder. Treatment of the material by severe plastic deformation resulted in the dissolution of the majority of oxides, as indicated by the decrease of the intensity of the appropriate x-ray peaks and the increase of the lattice parameter of the matrix based on Fe. The microstructure consisted of very small (0.05 μm) greatly disoriented

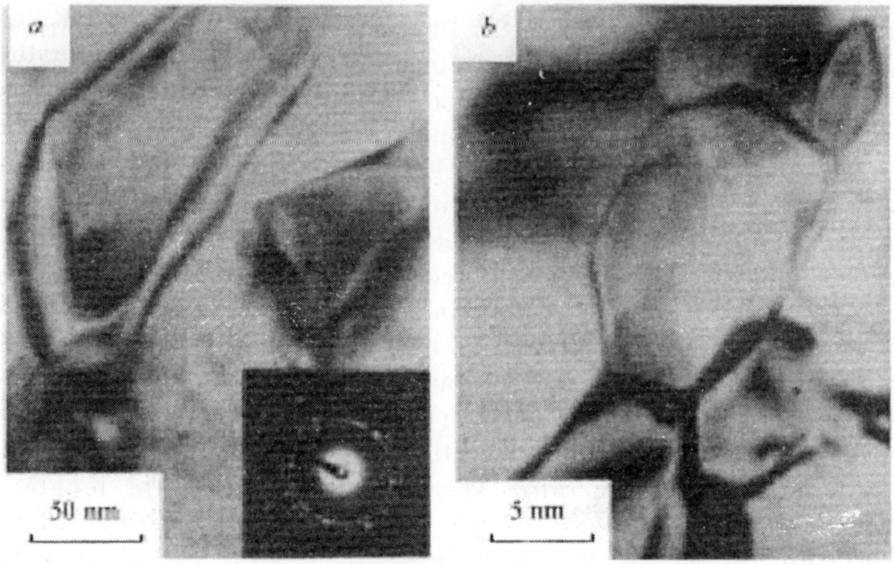

Fig. 4.18. Microstructure of Al–4% Cu–0.5% Zr alloy. *a*) after severe plastic deformation (true logarithmic strain $e = 7$), $T = 293$ K; *b*) after additional annealing at $T = 363$ K for 1 h 30 min.

grains with wide blurred boundaries (Fig. 4.19*a*). Annealing at T = 723 K resulted in some recovery in the structure. The grain size became equal to 0.1 μm and the grain boundaries became sharper. The microhardness, corresponding to this condition, slightly decreased in comparison with the condition immediately after severe plastic deformation. At T = 823 K examination showed the occurrence of recrystallisation, leading to the formation of clearly distinguished grains with a diameter of 0.2 μm and a small number (0.5%) of fine (<0.05 μm) oxide particles. Microhardness decreased more appreciably. At T = 873 K the specimens showed formation of oxide FeO whose volume fraction reached 18% (Fig. 4.19*b*). The precipitates appeared mainly at the grain boundaries of the matrix. At the same time, the lattice parameter decreased. The microhardness increased to the values higher than in the case of the material immediately after severe plastic deformation. At higher temperatures, examination showed the start of grain growth of the matrix and oxide particles, together with a larger decrease of microhardness.

Severe plastic torsional deformation under a pressure of 10 GPa of Fe–1.2% C pearlitic steel resulted in complete dissolution of cementite and the formation of a supersaturated solid solution of carbon in α-Fe [176]. The structure consisted of very fine

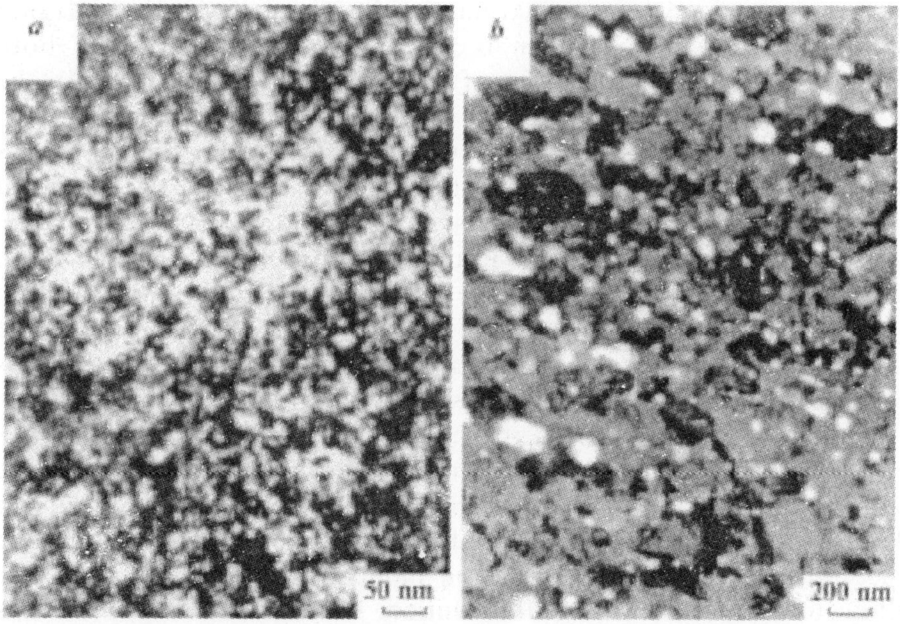

Fig. 4.19. Microstructure of Fe–O alloy. *a*) after severe plastic torsional deformation; *b*) after additional annealing at 873 K.

(0.02 μm) greatly disoriented nanograins. Annealing at T = 523 K led to to the formation of equiaxed grains free from dislocations and having the mean size of 0.05 μm. As shown by the examination of the temperature dependence of saturation magnetisation, cementite was not present in the material up to the given temperature. Heating to higher temperatures resulted in an increase in the degree of precipitation of cementite from supersaturated ferrite up to a temperature of 813 K, corresponding to the equilibrium composition of the alloy. The microhardness in the temperature range 473–673 K greatly decreased, indicating both the occurrence of recovery in the microstructure and the removal of surplus carbon from ferrite.

A different situation was found in the Cu–50 wt% Ag two-phase alloy in which both chemical elements were characterised by greatly restricted mutual solubility [181]. In the powders produced by gas condensation in an inert medium or by mechanical alloying, and in thick initial specimens, severe plastic torsional deformation resulted in the formation of a nanocrystalline (grain size 10–20 nm) supersaturated solid solution with a small amount of residual Cu and Ag phases. Annealing at a temperature of approximately 473 K resulted in the breakdown of the solid solution into grains of Cu and

Ag, characterised by approximately the same grain size. Higher temperatures resulted in an increase of the grain size of both components. The highest rate of this process was recorded in the monolithic material subjected to severe plastic deformation.

A very interesting behaviour was detected in heating of alloys of the Al–Fe system in which both chemical elements were almost completely insoluble in the solid state but severe plastic deformation resulted in the formation of a supersaturated solid solution [44]. Heating was accompanied by the breakdown of this solid solution (Fig. 4.20) with the formation of highly dispersed precipitates and the dislocations in the body of the grains and at the grains boundaries. This treatment increased the microhardness of the alloy.

Summarising the results, it may be concluded that the sequence of the processes, established in the course of evolution of the structure in heating of the pure metals, subjected to severe plastic deformation, is also obtained in alloys after identical treatment. The specific feature is the deformation-induced transition of two-phase alloys to a supersaturated solid solution. During annealing, there was a tendency for the reverse transition to the equilibrium condition by the precipitation and coalescence of the inclusions. In the investigated iron-based alloys (solid solutions to Fe–O and Fe–C) this tendency was detected in the final stage of evolution of the microstructure, i.e. during grain growth. In the Cu–Ag and Al–Fe

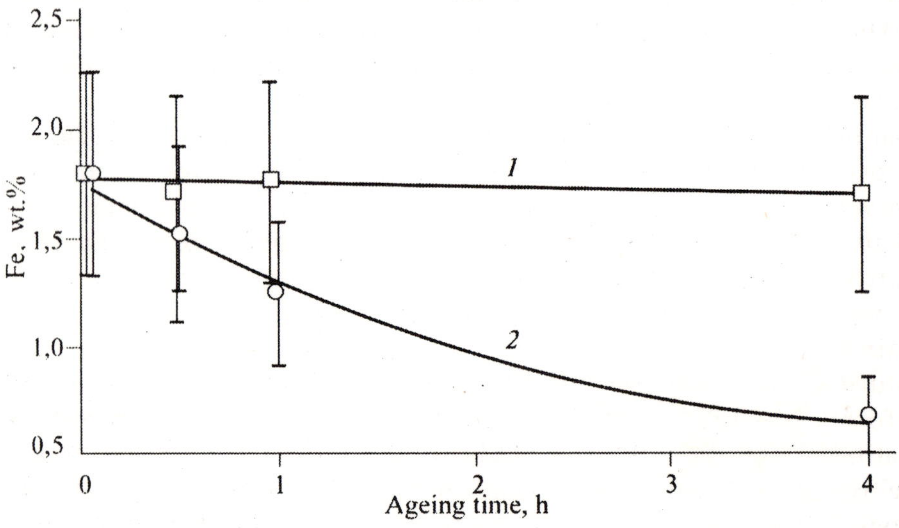

Fig. 4.20. Dependence of the concentration of iron in the solid solution of Al–11 wt% Fe alloy on ageing time, measured by the method of energy-dispersion spectroscopy with the beam diameter of 20 nm (*1*) and 0.5 μm (*2*). *1*) the local method, 2) the average method.

alloys, breakdown of the solid solution takes place prior to the start of grain growth and a high strength can be obtained in this case.

Intermetallics

The evolution of the microstructure in heat treatment of the intermetallics, subjected to severe plastic deformation, has been studied most extensively in the case of materials based on Ni_3Al [174, 177].

The structure of the intermetallic binary stoichiometric composition Ni_3Al, subjected to severe plastic deformation [211] consisted of very small equiaxed grains with features of the high level of internal stresses, as indicated by the complicated diffraction contrast on the electron micrographs and difficult to separate grain boundaries, having the form of diffuse and distorted boundaries. The mean size of the grains, determined on dark field images, was approximately 20–30 nm. A similar structure was also found in Ni_3Al subjected to severe plastic deformation in [177] but the grain size was slightly larger (50 nm). The condition after severe plastic deformation of Ni_3Al is also characterised by a high-level of electrical resistance, high internal stresses and high microhardness (Fig. 4.21). In addition to this, as indicated in Fig. 4.21, the results of x-ray diffraction analysis indicate the complete absence of the long-range order in this condition.

During annealing the microstructure undergoes a sequence of structural transformations, similar to those examined previously for nickel subjected to severe plastic deformation. However, the characteristic feature of Ni_3Al, subjected to severe plastic deformation, is that the long-range order starts to form in a narrow temperature range in the vicinity of T = 530 K, i.e. in the recovery stage (Fig. 4.21*e*). The ordering is not complete but a further increase of the parameter of the long-range order takes place only at considerably higher temperatures, close to 1300 K, when the grains grow to relatively large dimensions. Although the physical nature of disordering of the intermetallics in severe plastic deformation and subsequent reordering of the intermetallics in heating require further investigations, it is important to mention that the reordering in Ni_3Al is caused not by the recrystallisation processes but by the recovery processes associated with the rearrangement of the dislocation structure at the grain boundaries and in the body of the grains.

Identical results have obtained on the basis of the data presented in Fig. 4.22 [212]. In examination of the Ni_3Al intermetallic, alloyed with chromium and boron, and subjected to severe plastic torsional deformation at room temperature, attention was given to the structural changes in heating directly in the column of the electron microscope. At the same time, using differential scanning calorimetry, the heat generation in heating of the material was examined. As indicated by the results, the peak of heat generation

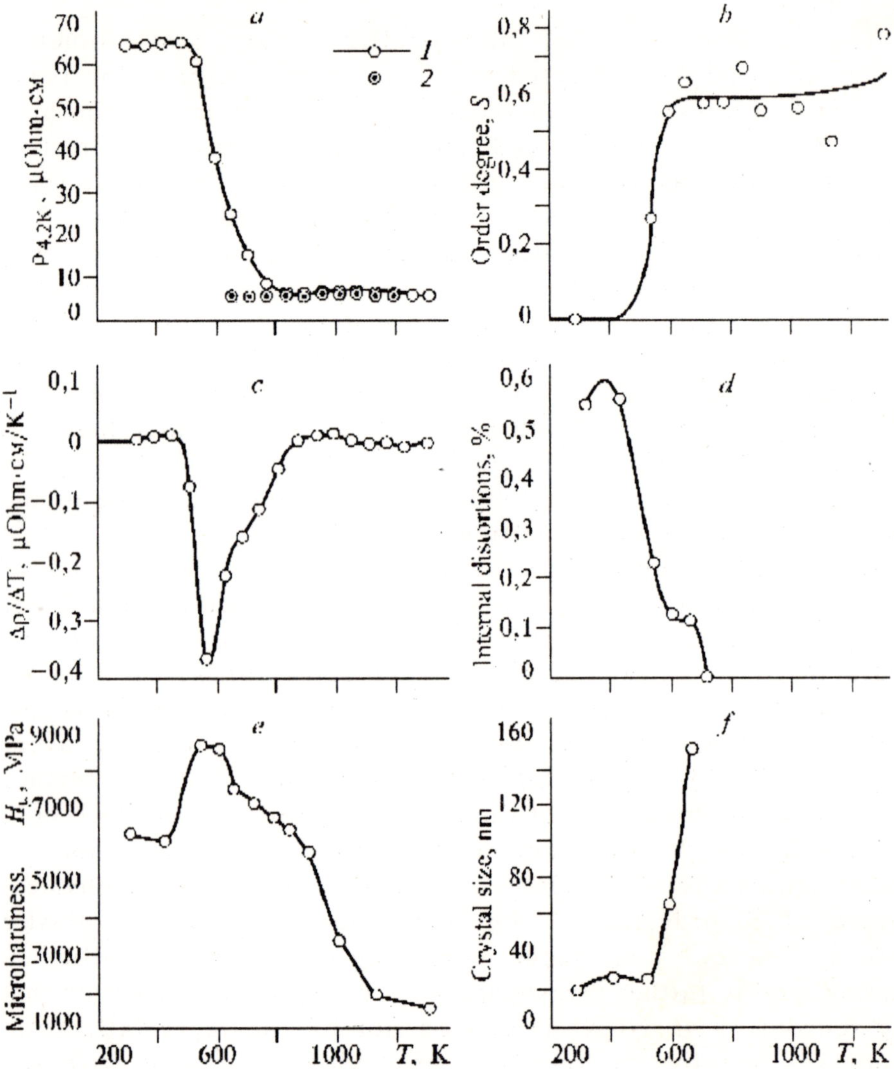

Fig. 4.21. Variation of electrical resistance (*a*), its temperature gradient (*b*), microhardness (*c*), the degree of order (*d*), internal stresses (*e*) and dimensions of crystals, obtained from the analysis of broadening of the x-ray peak (111), in relation to annealing temperature in nanostructured Ni_3Al (*f*).

is recorded at a temperature considerably lower than the start of rapid grain growth. The nature of this heat generation process is associated with the recovery processes and also with the start of reordering. It is important to know the high thermal stability of the nanostructured condition of this intermetallic which enables unique superplastic flow of this material to take place [213].

It is also important to note the study [214] in which attention was given to the effect of additions of boron (0.1 wt%) on the thermal stability of nanostructured Ni_3Al. The structure of this material with the addition of boron after severe plastic deformation was similar to the structure of pure Ni_3Al. Nevertheless, during annealing, the effect of boron was very strong. Recovery processes and changes of the properties in heating were also detected. However, they were greatly displaced in the direction of high temperatures. This effect is explained by the interaction of the boron atoms with dislocations and grain boundaries. Thus, the general sequence of the evolution of the structure in intermetallics based on Ni_3Al is similar to that established for pure metals and disordered alloys. However, the specific feature of this material is associated with the establishment of the long-range order already in early stages of the recovery process, i.e. in the redistribution and a decrease of the number of

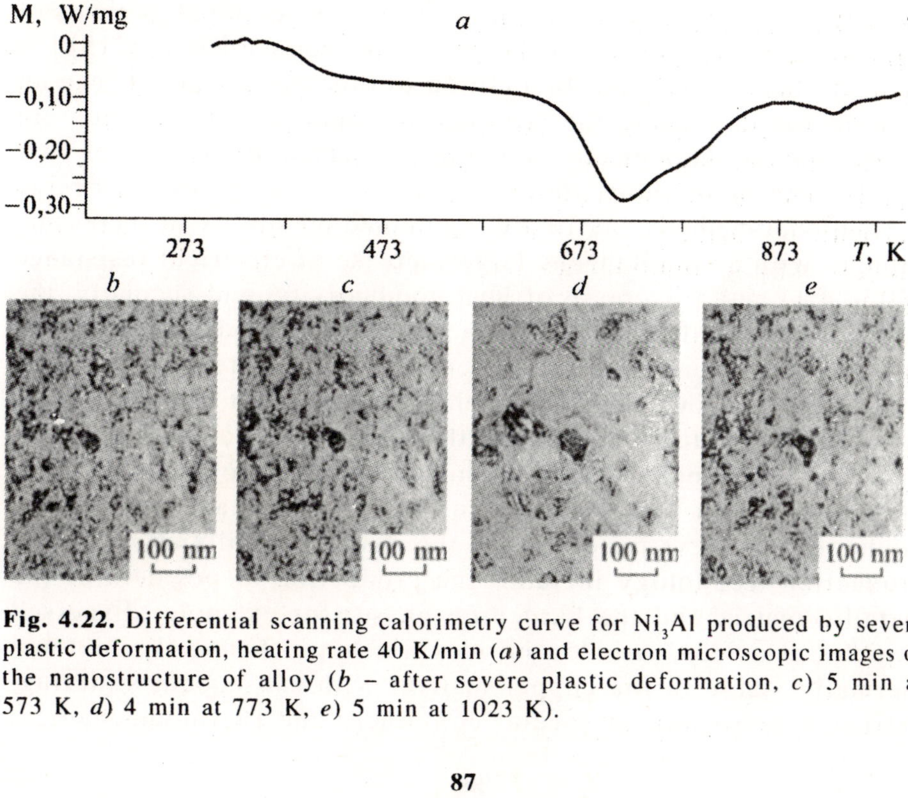

Fig. 4.22. Differential scanning calorimetry curve for Ni_3Al produced by severe plastic deformation, heating rate 40 K/min (*a*) and electron microscopic images of the nanostructure of alloy (*b* – after severe plastic deformation, *c*) 5 min at 573 K, *d*) 4 min at 773 K, *e*) 5 min at 1023 K).

dislocations. It has been assumed [178] that the direct reason for reordering is associated with mobile vacancies, formed as a result of the failure of various defects and dislocation loops, present in the deformed material.

Composites

It has already been mentioned that dispersion hardening is a promising method of increasing the thermal stability of the structure and mechanical properties of metallic materials. Several methods of dispersion hardening of metallic materials have been developed. They include the methods of dispersion hardening [215] (precipitation of a disperse secondary phase from a solid solution during ageing), the formation of composite materials [216], refining in a ball mill with subsequent consolidation of a mixture of powders [23], internal oxidation [217], etc. In the last decade, investigations were carried out into the possibilities of using the methods of dispersion hardening for development of ultrafine dispersed composites based on copper and silver characterised simultaneously by high electrical and heat conductivity with a sufficiently high strength.

It should be mentioned that the possibility of increasing the strength of copper and silver alloys by the conventional methods of solid-solution strengthening [218] and dispersion hardening [215] has basically been exhausted. In addition to this, these methods cannot be used for increasing the strength and stability of the structure whilst retaining high electrical and heat conductivity, typical of pure metals. The main shortcomings of these alloys, produced by the conventional methods, are firstly the limited possibility of increasing strength with a simultaneous large increase of electrical resistance (Table 4.1) and a decrease of heat conductivity, and secondly, the low thermal stability of the secondary phases, produced in ageing of the solid solution and, consequently, a large decrease of strength with a simultaneous decrease of electrical and heat conductivity at elevated temperatures because of the formation of a solid solution.

A large step forward in the solution of these problems has been the development of composite materials. The main methods of production of these materials are the powder metallurgy methods. Production technology includes the production of powders of the metallic matrix and the hardening phase (for example, dispersed particles of oxides, carbides, nitrides which are chemically inert and thermally stable in the copper matrix) with subsequent sintering, extrusion or rolling in special containers and thermomechanical treatment. This technological cycle results in the formation of a compacted pore-free material with the uniform distribution of the hardening phase in the volume of the dispersed particles.

Significant advances in the development of high-strength composite materials based on copper and silver, retaining high electrical and heat conductivity, have been made as a result of the development of methods of preparation of alloyed dispersed powders such as internal oxidation and milling in a ball mill (Table 4.1). However, like the conventional methods of hard-alloy strengthening or dispersion hardening, the possibilities of development of composite materials in respect of many parameters may be regarded as realised.

Taking the above considerations into account, the promising direction in the solution of this problem is is the application of the methods of severe plastic deformation for the development of nanocomposites with a metallic base. In the treatment of the composites, produced by the method of internal oxidation or

Table 4.1. Mechanical and electrical properties of copper and copper-based alloys

Material	Strengthening method	Yield limit	Tensile strength	Plasticity, %	Ratio of conductivities of strengthened materials and pure copper
		MPa			
Recrystallised copper	–	60	230	50	1 [217]
Cu–Al	Solid-solution strengthenning	130	330	55	3 [219]
Cu–Zn	As above	250	420	70	6 [220]
Cu–Be	Dispersion hardenning	300	500	35	4 [220]
Cu– 0.5 wt.% Al_2O_3	Internal oxidation	355	425	13	1.25 [220]
Cu	SPD by torsion	370	390	2.5	1.4 [180]
Cu	SPD by ECA pressing	410	425	12	1.4 [180]
Cu– 0.5 wt.% Al_2O_3	Internal oxidation + SPD in torsion	475	510	3	1.28 [187]

consultation of a mixture of ultrafine dispersion powders, one of the methods of severe plastic deformation makes it possible to produce in the metallic matrix a homogeneous grain structure with the grain size of 0.1–0.3 μm which increases the strength whilst retaining satisfactory plasticity (Table 4.1). The grain boundaries of the composites, produced under the effect of severe plastic deformation, are characterised by a specific nonequilibrium structure which also contributes to strengthening. The thermal stability of the structure of this composite is associated with the presence of nanosized particles (10–50 nm), uniformly distributed in the matrix.

The following is a review of the results of investigations of the structure, the chemical properties and thermal stability of the composite materials based on copper and silver, produced by different methods and subjected to different types of treatment, including severe plastic deformation, and the results of investigations carried out by the authors of this book are also presented.

The thermal stability of the composites, produced by consolidation of a mixture of ultrafine dispersed powders, has been investigated by many authors [221–228]. In particular, the authors of [221] investigated the structure of dispersion-hardened composites Cu–3 vol% TiC and $CuTi_4$–10 vol% TiC which in the initial condition (after consolidation of the powder mixture and hot extrusion) have the mean grain size of the matrix of ~0.5 μm and of the TiC disperse phase 10–30 nm. The particles of the strengthening phase were distributed mainly at the grain boundaries. It has been established that annealing at a temperature of 573–973 K of the $CuTi_4$–10 vol% TiC alloy increases the grain size of the matrix and also the grain size of the disperse TiC phase. However, in the Cu–3vol% TiC alloy after annealing in these conditions examination showed only grain growth in the metallic matrix (Table 4.2). The high thermal stability of the structure of the Cu–3vol% TiC composite has also been reported by the authors of [187] in the examination of the stability of the structure of this composite in relation to the effect of temperature and deformation (Table 4.2). The Cu–3 vol% TiC composite was investigated into two conditions: immediately after hot extrusion and subsequent severe plastic torsional deformation under high pressure (true logarithmic strain ~7).

Examination by the method of extraction replicas showed that in both types of specimens the particles of the strengthening phase TiC with the size of approximately 10 mm are distributed uniformly

Table 4.2. Temperature of the start of the process of recrystallisation of nanostructured copper and ultrafine-grained copper-based composites

Material	Method of producing structure	T, K
Cu–0.5 wt.% Al_2O_3	Extrusion	673 [187]
Cu–0.5 wt.% Al_2O_3	Extrusion + Cold rolling to 70%	673 [187]
Cu–0.5 wt.% Al_2O_3	Extrusion + torsional SPD	673 [187]
Cu–0.9 wt.% TiC	As above	>873 [187]
Cu–3 vol.% TiC	Extrusion	573 [221]
Nanostructured copped	Torsional SPD	448 [229]

in the matrix. The distribution of the larger particles (mean size 350 nm) and the volume is nonuniform. There are long areas with and without these particles. The grain size of the copper matrix of the consolidated composite in the initial condition was 0.5 μm, and that of the composite subjected to severe plastic deformation (nanostructured composite) was 0.3 μm.

Annealing at a temperature of 673–1073 K does not influence of the size or distribution of the matrix of the particles of the strengthening phase in both types of specimens. The grain structure of the copper matrix in both types of specimens is the thermally stable up to a temperature of a candid and 873 K (Table 4.2). Annealing at higher temperatures increases the grain size; after annealing for 1 h at a temperature of 1073 K, the grain size reaches 1.5–2 μm. Thus, the formation of the nanostructure in the Cu–3 vol% TiC composite under the effect of severe plastic deformation in torsion has no effect on the temperature of the start of recrystallisation of its copper matrix.

Detailed examination of the stability of the structure of the composites, produced by consolidation of the internally oxidised mixture of powders with subsequent hot extrusion, in relation to the effect of temperature and deformation was carried out by the authors on the example of the composites Cu–0.5 wt% Al_2O_3 [187, 228–230). Three conditions of the composites were investigated, after extrusion (condition 1), after extrusion and subsequent deformation to 70% in the rolling (condition 2), after extrusion and subsequent torsional deformation by the 'pressure + shear' scheme (condition 3–nanostructured composite) (Table 4.2).

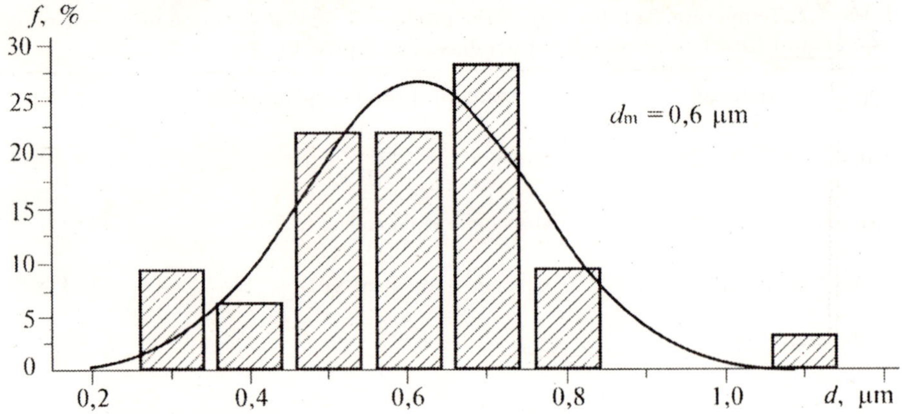

Fig. 4.23. The size distribution of the grains of the copper matrix of the Cu–0.5 wt% Al_2O_3 composite.

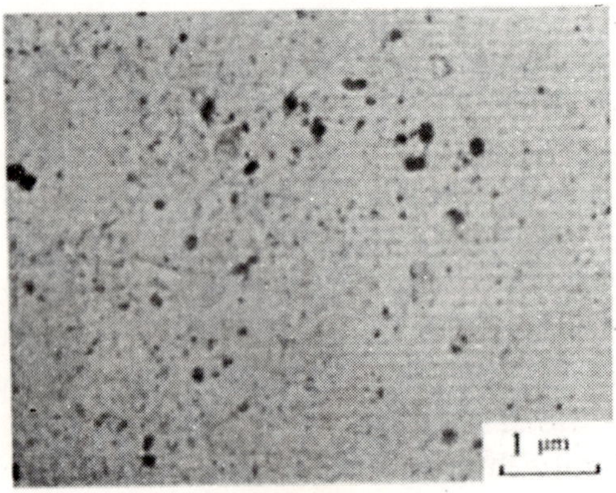

Fig. 4.24. The image of a carbon replica with extraction of Al_2O_3 particles (Cu–0.5 wt% Al_2O_3 composite).

The results of electron microscopic and metallographic examination show that after extrusion the Cu–0.5 wt% Al_2O_3 composite has a fine-grained structure, the grain size 0.5–1 μm (the size distribution of the grains is shown in Fig. 4.23). The particles of the strengthening phase Al_2O_3 with the size smaller than 20 nm are uniformly distributed in the volume of the specimen (Fig. 4.24). The structure also contains larger particles of the Al_2O_3 phase with the size reaching 300 nm (Figs. 4.24 and 4.25). In addition to this, there were individual precipitates of the Al_2O_3 phase with the size of ~10 μm. Subsequent torsional deformation reduced the mean grain size of the copper matrix to 0.3 μm (Fig. 4.26). The dimensions and distribution of the Al_2O_3 particles did not change. Deformation by rolling had also no effect on the size and distribution of the Al_2O_3 particles.

After annealing at a temperature of 673 K for 1 h, the specimens of the Cu–0.5 wt% Al_2O_3 composites showed, irrespective of preliminary treatment, the formation of the first recrystallised areas with sharp boundaries with non-recrystallised

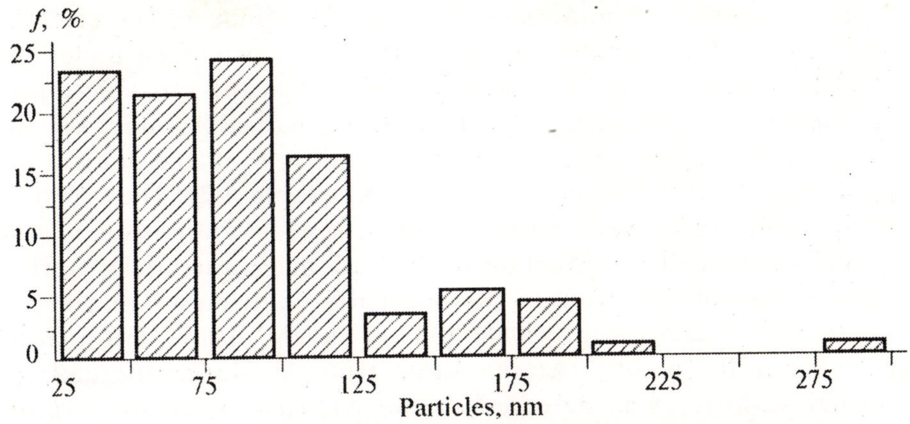

Fig. 4.25. The size distribution of Al_2O_3 particles in the Cu–0.5 wt% Al_2O_3 composite.

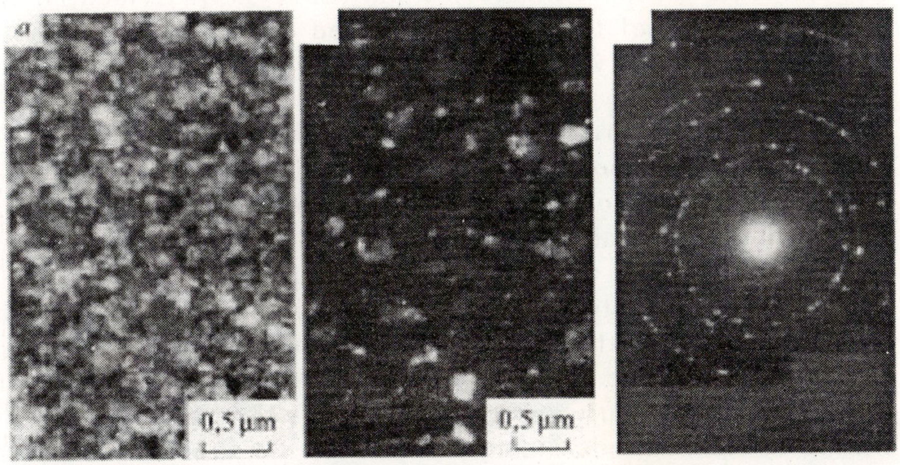

Fig. 4.26. Bright field (*a*) and dark field (*b*) images of the structure and the microdiffraction pattern (*c*) of the Cu–0.5 wt% Al_2O_3 composite, produced by severe plastic torsional deformation.

areas and a high dislocation density. Examination by the replica method shows that the area of the recrystallised regions of the surface does not exceed 5–6%. After annealing at a temperature of 873 K for 1 h, the area of the recrystallised parts equalled 90%. The recrystallised grains showed a large number of annealing twins. The grain size reached 2–3 μm. After annealing at a temperature of 1073 K for 1 h, complete recrystallisation of the copper matrix of the Cu–0.5 wt% Al_2O_3 nanostructured composite took place. The mean grain size was 2.5 μm. Annealing for 1 h at a temperature of 1273 K increased the mean grain size of the copper matrix to 3.5 μm.

Comparison of the results of examination of the thermal stability of the structure of pure nanostructured copper [231] and the Cu–0.5 wt% Al_2O_3 nanostructured composite, produced under the effect of severe plastic deformation shows that the introduction to the copper matrix of the nanosized Al_2O_3 particles increases the temperature of the start of recrystallisation of the matrix by approximately 200 K (from 478 to 673 K) (Table 4.2). These results are in good agreement with the results of examination of the thermal stability of the structure of the bulk nanostructured materials produced by consolidation of the internally oxidised mixture of the nanosized powders using the methods of severe plastic deformation carried out in [6, 40]. At the same time, no large increase of thermal stability was detected in the Al+10% Al_2O_3 + 15% SiC composites, containing a large fraction of the ceramic particles, but with the size of approximately 1 μm [232, 233]. Evidently, the thermal stability of the ultrafine-grained composite materials is determined primarily by the dispersion and morphology of the ceramic particles.

The process of recrystallisation in the Cu–0.5 wt% Al_2O_3 nanostructured composite is slow and the increase of temperature from 673 to 873 K does not the result in complete recrystallisation of the material. The absence of polygonisation in non-recrystallised areas is evident. It is well known that polygonisation in, for example, dispersion-strengthened TD nickel is a mechanism delaying recrystallisation in the system [234]. (Evidently, the absence of polygonisation in the Cu–0.5 wt% Al_2O_3 system is associated with the low stacking fault energy of copper which complicates the sliding of dislocations during annealing). On the basis of these results it may be concluded that the main mechanism of recrystallisation in the investigated nanostructured composite is the nucleation and growth of new grains [187]. It should be mentioned that the main mechanism of recrystallisation of pure copper, subjected to severe plastic deformation, is the migration of of equilibrium grain boundaries [235].

Examination of the structure-sensitive properties (microhardness and specific electrical resistance) of the Cu–0.5 wt% Al_2O_3 nanostructured composite in the process of evolution of the structure and comparison of the results with the values obtained for pure copper show that the relationships of the variation of these properties during these treatments differ.

Figure 4.27 shows the results of measurements at 77 K of the electrical resistance of the Cu–0.5 wt% Al_2O_3 nanostructured composite in relation to annealing temperature. It may be seen that

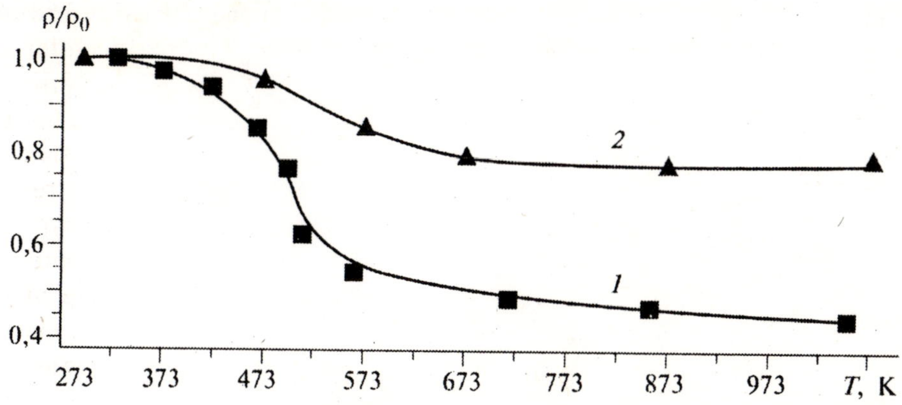

Fig. 4.27. Dependence of the specific electrical resistivity of nanostructured copper (*1*) and Cu–0.5 wt% Al_2O_3 composite (*2*) on preliminary annealing temperature.

the main variation of (by approximately 20%) takes place after annealing at a temperature of 573 and 673K. At annealing temperatures higher than 673 K, the value of ρ does not change. In pure nanostructured copper, subjected to torsional deformation, the main decrease of ρ is recorded after annealing in the temperature range 473–573 K [231]. The authors of [231] link the observed decrease of this parameter in pure copper with the increase of the grain size from 0.2 to 13 μm. The large decrease of the value of ρ of the investigated nanostructured composite after annealing in the temperature range 573–673 K is evidently associated with the recovery of the structure of the grain boundaries (transition of the grain boundaries to the equilibrium condition) because according to the results of examination of the structure there is no significant grain growth after annealing in these conditions.

An important special feature of the ultrafine copper-based dispersed composites is the high thermal stability of microhardness (H_μ) (Fig. 4.28) and the Vickers hardness (Table 4.3) [224, 227] in comparison with pure coarse-grained [227] and nanostructured copper [183]. For example, in the initial condition and after annealing in the temperature range 673–873 K, the microhardness of the nanostructured composite Cu–0.5 wt% Al_2O_3 was 1.60–1.67 GPa. After annealing at higher temperatures (923–1073 K) the microhardness decreases to 1.40 GPa but still remains twice that of recrystallised copper (0.27 GPa) (Fig. 4.28).

The variation of microhardness of other investigated nanostructured composites after preliminary annealing shows the same

Table 4.3. Dependence of Vickers hardness of ultrafine-grained composites based on copper (VH 10) on preliminary annealing temperature

Material	Annealing temperature, K				Author
Pure copper	115	36	32	20	[227]
Cu–3 vol.% Al_2O_3	220	210	120	100	[227]
Cu–10 vol.% Al_2O_3	255	215	155	105	[227]
Cu–1 vol.% Al_2O_3	80	80	73	68	[224]
Cu–3 vol.% Al_2O_3	92	95	88	80	[224]

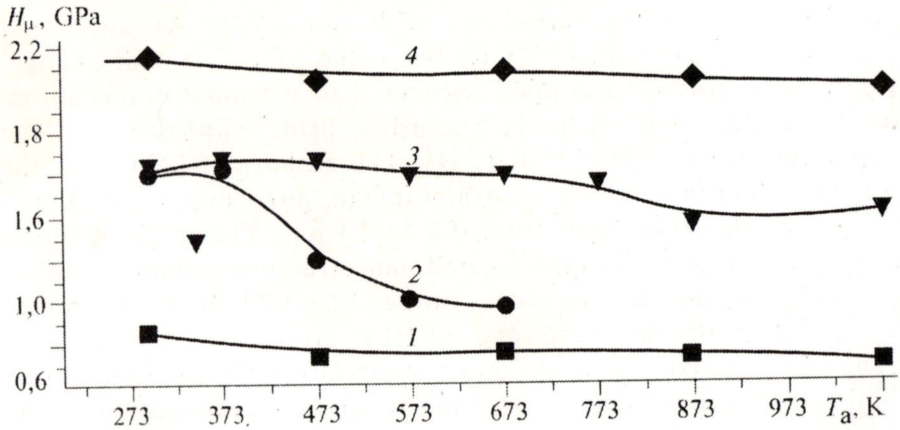

Fig. 4.28. Dependence of microhardness on preliminary annealing temperature for coarse-grained (*1*) and nanostructured (*b*) copper and for nanostructured composites Cu–0.5 wt% Al_2O_3 (*3*) and Cu–0.9 wt% TiC (*4*).

trend (Fig. 4.28). At the same time, the microhardness of pure nanostructured copper already decreases rapidly after annealing at a temperature of 473 K [183].

Summing up to the results presented in this section, it may be seen that the evolution of the structure in heating of different nanostructured materials has a number of common features. The process consists of several stages which, however, may overlap:

– the redistribution and a decrease of the number of dislocations existing in the grains of the material subjected to severe plastic deformation;

– the redistribution of the dislocations in the nonequilibrium boundaries of the grains, formed during severe plastic deformation resulting in the formation of high-angle grain boundaries with a thickness comparable with the size of the atoms. The simultaneous decrease of the long-range stress fields and elastic distortions of the crystal lattice as a result of structural recovery of nonequilibrium grain boundaries is evident. This is accompanied by the formation of a polycrystalline structure with the grains in the micron size range. There is no nucleation stage. Usually, this stage is not accompanied by the migration of the grain boundaries and the mechanism corresponds to *in situ* recrystallisation. However, if some grain boundaries undergo recovery faster than others and are capable of migration, absorbing adjacent grains, the mechanism corresponds to conventional recrystallisation.

– grain coarsening during heating. If after recovery the structure contains individual nonequilibrium grain boundaries, the process of recrystallisation may be accompanied by anomalous grain growth. The temperatures resulting in some processes of evolution of the structure depend on several factors, namely:

– the nature of the investigated material and its chemical composition (main elements and, in particular, alloying additions and inclusions which may greatly delay evolution);

– the method of parameters of severe plastic deformation (equal-channel pressing, torsional or combination of the processes) affecting the structure, formed immediately after deformation, and consequently, through the structure, the sequence of the processes, taking place during subsequent annealing.

In alloys and intermetallics, the changes in the defective structure are accompanied by additional phenomena, associated with the redistribution of the atoms of different chemical elements. The redistribution of the atoms, resulting in disordering and phase transformations, is caused by severe plastic deformation. During annealing, there is a tendency for the transition of the material to the equilibrium condition. This may be accompanied by:

– local atom rearrangement encountered in the restoration of the long-range order (no investigations were not carried out for short-range ordering);

– the displacement of the atoms over large distances essential for phase transformations, such as the precipitation of dislocations and coalescence of the second phase in the solid solution which was supersaturated during severe plastic deformation (the position in the sequence of structural evolution in which these phenomena take place, is associated with the redistribution of the atoms and depends on the type of alloy).

Until recently, the mutual effect of these two mechanisms of evolution of the structure (the variation of the defective structure of the crystal lattice and the variation of the distribution of atoms of different chemical elements) during annealing of the defined alloys and intermetallics has been studied insufficiently. Evidently, the examination of the mutual effect of these mechanisms and also the examination of the relationship between the structural changes and changes of the properties occupies an important position in further investigations, aimed both at understanding of the fundamental processes taking place during annealing of material subjected to severe plastic deformation and at examination of the thermal stability of sub-microcrystalline materials during industrial application.

4.2. DEFORMATION BEHAVIOUR OF NANOSTRUCTURED MATERIALS UNDER ACTIVE TENSILE LOADING

The extensive practical application of the bulk nanostructured materials whose ultrafine-grained structure is produced by the effect of severe plastic deformation, as structural materials is associated mainly with the problem of the instability of the nonequilibrium structure of these materials. The instability, associated with the high energy stored in severe plastic deformation, is the reason for intensive softening at relatively low temperatures. These temperatures for several nanostructured metals, as already mentioned, are close to room temperature.

Dispersion strenthening is an effective method of increasing the thermal stability of the mechanical properties of metals and alloys. Recently, investigations have been carried out to develop scientific fundamentals of the production of composites whose matrix is treated by different methods to produce the nanostructured condition (grain size 0.1–0.3 μm) in addition to strengthening by the particles of high-stability oxide, carbide and other types of phases whose size is in the nanorange (10–15 nm). The low solubility of the strengthening phases and the high resistance of the phases to the coalescence processes make it possible to assume that the nanocomposites have not only a higher level of mechanical properties but also higher stability in the conditions of the effect of temperature and loading.

Below are the results of experimental investigations of the mechanical properties of nanostructured metals and nanocomposites based on these metals, produced under the effect of severe plastic deformation. Special attention is given to examination of the effect of heat treatment and the diffusion processes, taking place during heat treatment, on the deformation behaviour of these materials in different types of loading.

Pure metals

It is well known that the formation of a fine-grained structure is one of the methods of increasing strength whilst retaining satisfactory plasticity of polycrystalline materials [206]. However, a decrease of the grain size to the nanorange is not always accompanied by an increase of the strength characteristics. For example, using MA8 alloy, Musalimov and Valiev showed that in tensile loading nanostructured materials in the grain size range smaller than 500 nm the material may soften [236]. In the nanorange, this material does not show any significant strain hardening in contrast to the alloy with the grain size of ~10 μm [237]. Panin et al [238, 239], when examining the deformation behaviour of nanostructured copper (*d* ~ 200–500 nm) in tensile loading, found that the yield limit of the material at room temperature ($\sigma_{0.2}$) is almost 7 times higher than the yield limit of the coarse-grained condition (*d* ~ 40 μm). As in the case of MA8 alloy, nanostructured copper does not show strain hardening, characteristic of conventional polycrystals. The identical behaviour in tensile loading was reported by the authors for titanium [198], copper [240] and nickel (Fig. 4.29*a*, Table 4.4). The results show that preliminary pre-recrystallisation annealing which does not

Table 4.4. Yield limit ($\sigma_{0.2}$) and tensile strength (σ_B), strain to fracture (Δ) and microhardness of nanostructured copper, nickel and titanium, produced by severe plastic deformation using equal-channel angular pressing

Material	$\sigma_{0.2}$, MPa	σ_B, MPa	Δ, %	H_μ, GPa
Copper	405	426	9.1	1.52
Nickel	989	1018	6.1	2.64
Titanium	654	716	8.9	2.3

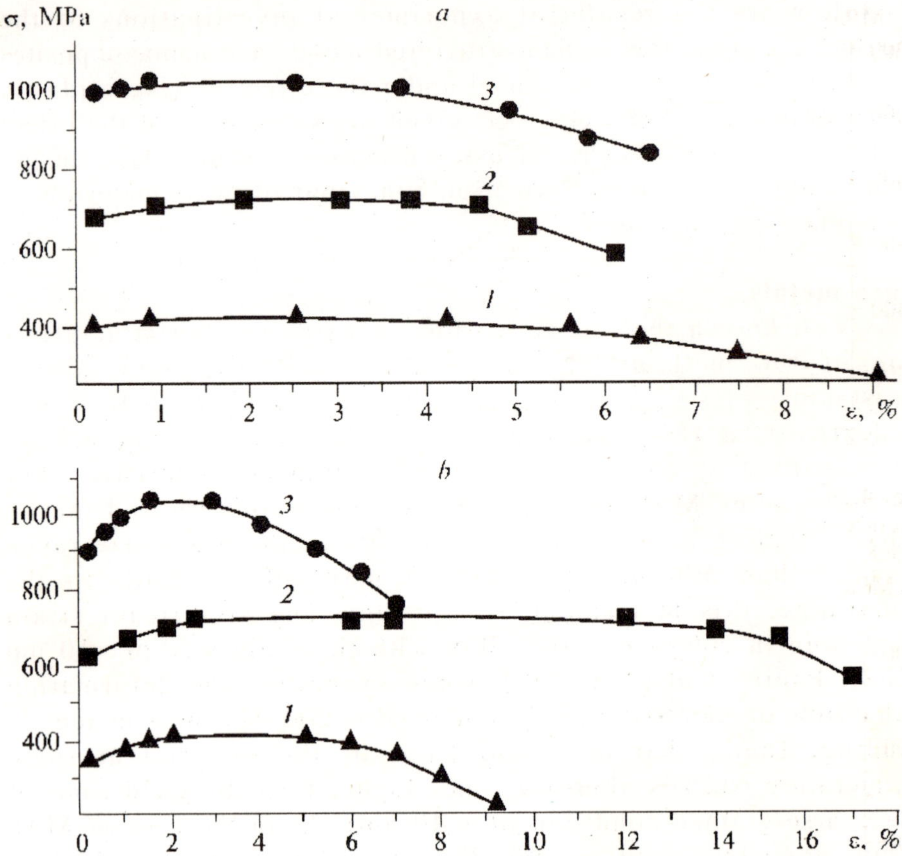

Fig. 4.29. Flow curves of specimens of copper (1), titanium (2) and nickel (3) in the nanostructured condition (*a*) and after low-temperature annealing (*b*) (Ti – 473 K, 1 h; Ni – 398 K 1 h; Cu– 353 K, 1 h); T_t = 293 K.

Table 4.5. Yield limit ($\sigma_{0.2}$) and tensile strength (σ_B), strain to fracture (Δ) and microhardness of nanostructured copper, nickel and titanium after equal-channel angular pressing and subsequent low temperature annealing

Material	$\sigma_{0,2}$ (MPa)	σ_B (MPa)	Δ, %	H_μ, GPa	*T*, K
Copper	350	410	9.2	1.43	353
Nickel	952	1050	4.2	2.54	398
Titanium	619	719	17.2	2.48	573

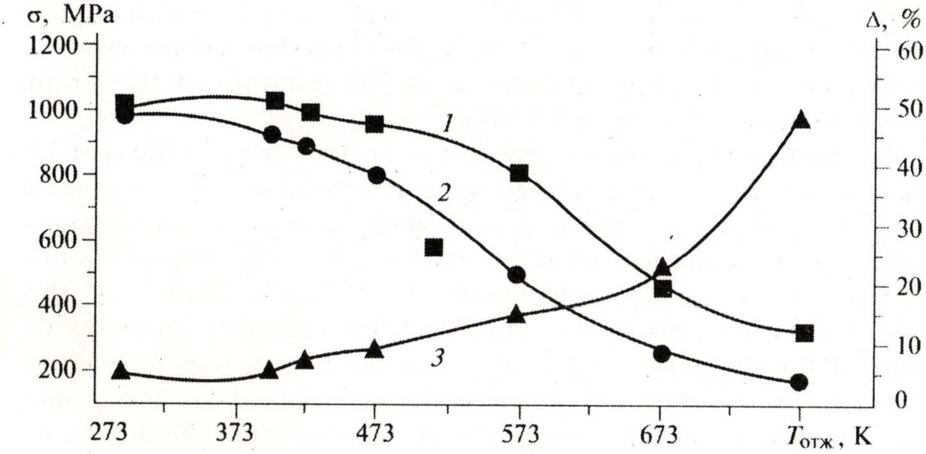

Fig. 4.30. Dependence of tensile strength (*1*) and yield limit (*2*) and strain to fracture (*3*) of nanostructured nickel on preliminary annealing temperature, T_t = 273 K.

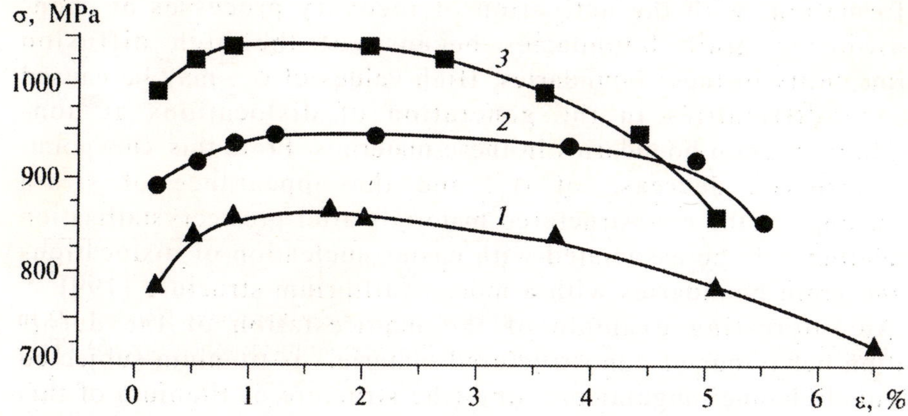

Fig. 4.31. Flow curves of specimens of nanostructured nickel after preliminary annealing at a 398 (1), 473 (2), 523 K (3); T_t = 293 K.

change the grain size of the investigated nanostructured materials (for example, annealing of copper at a temperature of 353 K, Ni at 398 K, Ti at 573 K), nevertheless results in a decrease of $\sigma_{0.2}$ (Table 4.5) and the appearance of the effect of strain hardening (Fig. 4.29*b*). The yield limit remains almost constant in this case (Table 4.5).

The detailed examination of the dependence of the mechanical properties of nanostructured metals on annealing temperature, carried out on nickel after equal-channel angular pressing, showed that the increase of the temperature of preliminary pre-recrystallisation annealing results in a monotonic decrease of the yield and tensile strength with a simultaneous increase of strain to fracture (Fig. 4.30). The flow curves (stress-strain), Fig. 4.31, show an increase of the duration of the strain hardening stage of strain hardening.

The previously examined difference in the strain behaviour of coarse-grained and investigated nanostructured metals, produced under the effect of severe plastic deformation, is associated, according to the authors of [239], with the fact that the relaxation time of atomic rearrangement in the defective matrix of the nanostructured metals is very short and, consequently, the stresses required for the generation of dislocations during tensile loading cannot form. For this reason, the dislocation density in the grains does not increase and, consequently, there is no strain hardening of the material.

The authors of [199] link the slight strain hardening of nanostructured materials, produced by the effect of severe plastic deformation, with the activation of recovery processes at non-equilibrium grain boundaries because of the high diffusion permeability of these boundaries. High values of $\sigma_{0.2}$ may be caused by the difficulties in the generation of dislocations at non-equilibrium grain boundaries in these materials. From this viewpoint, the observed decrease of $\sigma_{0.2}$ and the appearance of strain hardening for the nanostructured materials after pre-recrystallisation annealing may be associated with easier nucleation of dislocations at the grain boundaries with a more equilibrium structure [199].

An interesting example of the manifestation of the deformation behaviour of nanostructured materials is titanium subjected to equal-channel angular pressing (the structure of titanium of this type has been described in detail in section 4.1). Comparative investigations of the deformation behaviour of nanostructured and

Table 4.6. Mechanical properties of VT1-0 titanium, subjected to equal-channel angular pressing, in the longitudinal and transverse sections in relation to the pressing direction at room temperature (strain rate $3.3 \cdot 10^{-3}$ s^{-1}).

Material	σ_B	$\sigma_{0.2}$	Δ, %	H_μ, GPa
	MPa			
NS of titanium, longitudinal section	716	654	8.9	2.3
NS of titanium, transverse section	809	769	3.0	2.27

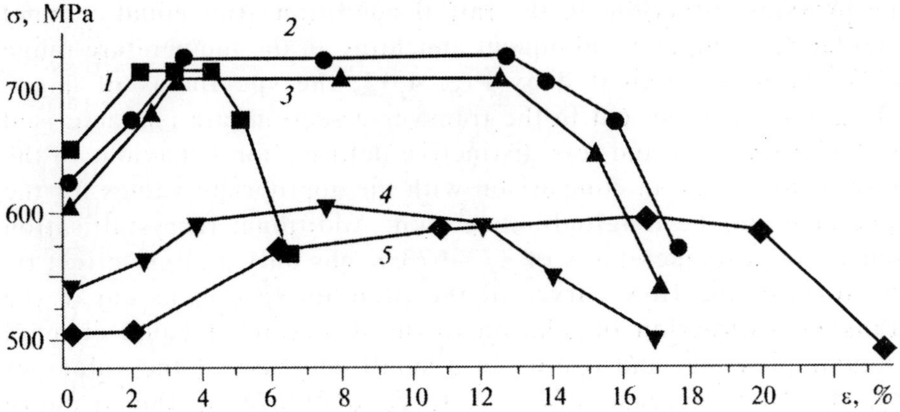

Fig. 4.32. Flow curves of the specimens of nanostructured titanium, cut out in the direction parallel to the axis of equal-channel angular pressing: T_t = 293 K and $\dot{\varepsilon}$ = $3.3 \cdot 10^{-3}$ s^{-1}. *1*) initial; *2*) at T = 473 K, *3*) 573 K, *4*) 673 K, *5*) 773 K.

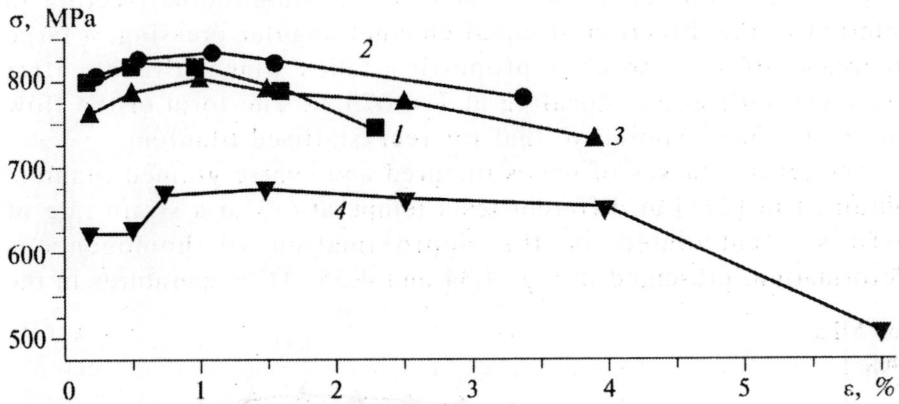

Fig. 4.33. Flow curves of the specimens of nanostructured titanium, cut out in the direction normal to the axis of equal-channel angular pressing; T_t = 293 K and $\dot{\varepsilon}$ = $3.3 \cdot 10^{-3}$ s^{-1}) initial; *2*) at T = 473 K, *3*) 673 K, *4*) 773 K.

coarse-grained (d = 10 μm) titanium at temperatures below the recrystallisation temperature of nanostructured titanium were carried out in [241].

The results of tensile tests show that the mechanical properties of nanostructured titanium in the longitudinal and transverse cross sections in relation to the pressing direction greatly differ (Table 4.6).

Figures 4.32 and 4.33 show the flow curves in tensile loading the specimens of nanostructured titanium, cut out in the longitudinal (Fig. 4.32) and transverse (Fig. 4.33) cross sections in relation to the pressing direction, in the initial condition after equal-channel angular pressing and subsequent annealing in the temperature range 473–773 K As indicated by Fig. 4.33, the specimens of nano-structured titanium, cut in the transverse section, are characterised by lower ductility and less distinctive deformation behaviour in the plastic flow stage in comparison with the appropriate values for the specimens of the longitudinal section. Additional recrystallisation annealing at temperatures of 473–673 K has only a slight effect on the form of the flow curves of the titanium specimens cut in the transverse direction in relation to the direction of equal-channel angular pressing, and results in a small decrease of the values of σ_B, $\sigma_{0.2}$ and increase of Δ. A large decrease of the strength properties and the change in the form of the flow curves of the investigated specimens of titanium are detected only after annealing at a temperature of 773 K, i.e. after recrystallisation of the structure.

In the titanium specimens, cut from the longitudinal section in relation to the direction of equal-channel angular pressing, a large decrease of the strength properties takes place already after pre-recrystallisation annealing at T = 673 K. The form of the flow curves becomes similar to that for recrystallised titanium.

The tensile curves of nanostructured and coarse-grained titanium obtained in [241] at different tests temperatures at a strain rate of $3\cdot10^{-5}s^{-1}$, calculated in the approximation of homogeneous deformation, presented in Fig. 4.34 and 4.35. At temperatures in the

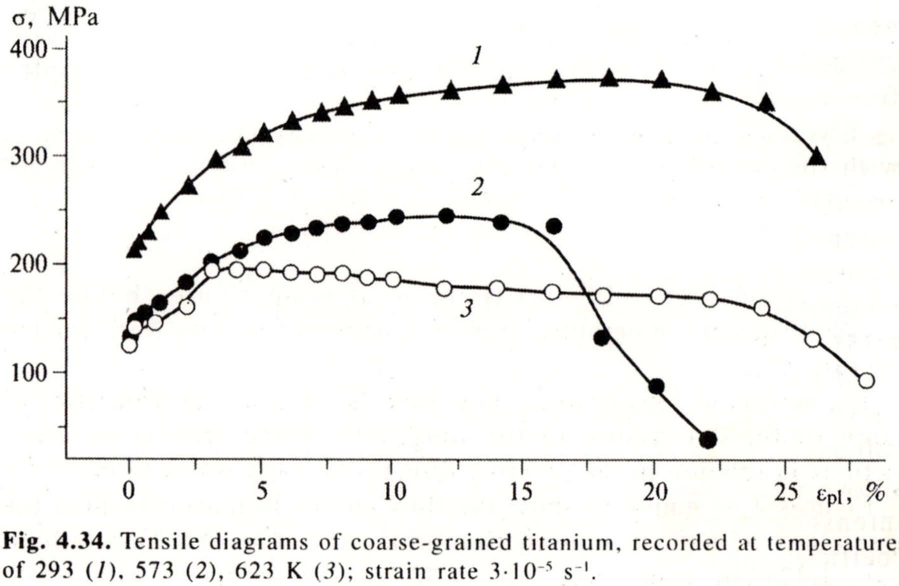

Fig. 4.34. Tensile diagrams of coarse-grained titanium, recorded at temperatures of 293 (*1*), 573 (2), 623 K (*3*); strain rate $3\cdot10^{-5}$ s^{-1}.

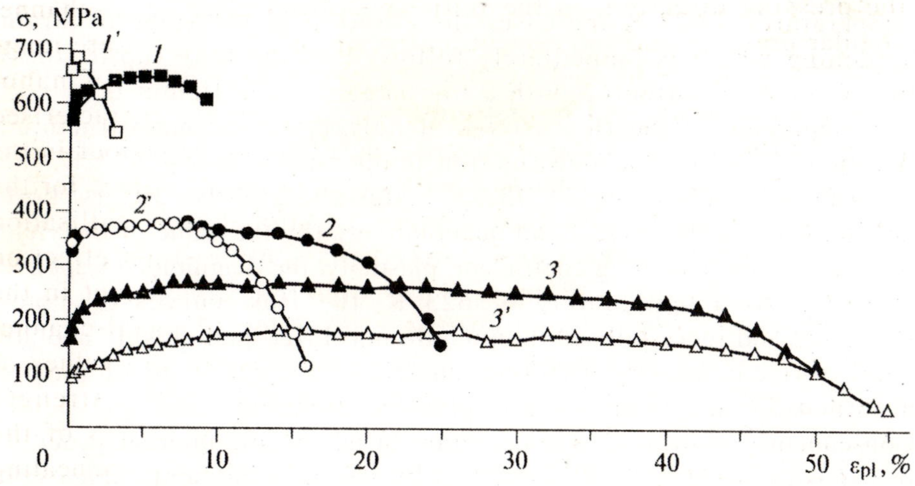

Fig. 4.35. Tensile diagrams of nanostructured titanium in the longitudinal (1–3) and transverse (1'–3') directions in relation to the pressing direction at temperatures of 293 (*1*, *1'*), 573 (2, 2'), 623 (*3*, *3'*); strain rate $3\cdot10^{-5}$ s^{-1}.

range from 293 to 573 K, the flow curves of coarse-grained titanium have the usual parabolic form with a large decrease of the strain hardening coefficient $\theta = d\sigma/d\varepsilon_{pl}$ with an increase of plastic strain (Fig. 4.34). The increase of deformation temperature to 623 K changes the form of the flow curves. The duration of the stage of strain hardening decreases from 15 to 4% and is followed by a linear stage (to 24%) of strain softening with a low and negative (~100 MPa) coefficient θ. The yield limit of coarse-grained titanium is almost halved with an increase of temperature from room to 623 K, and the general plasticity changes only slightly.

The flow curves of the specimens of nanostructured titanium with the tensile axis along and across the axis of equal-channel angular pressing in the temperature range 293–623 K are also parabolic. In this case, the nature of deformation behaviour of nanostructured titanium depends more appreciably on deformation temperature than in the case of polycrystalline titanium (Fig. 4.35). At room temperature, the nanostructured titanium has a high yield limit $\sigma_{0.2}$ (up to 700 MPa) and low plasticity. On the flow curve, the short stage of strain hardening is replaced by rapid softening. At T = 573 K, the flow curves of nanostructured titanium shows clearly several stages of strain hardening: the stage of intensive hardening is replaced by the linear stage with a low coefficient of strain hardening which is followed by two linear stages of softening with higher negative coefficients θ. The increase of temperature to 623 K increases the duration of the stage of strain hardening which is immediately followed by the long (up to 40%) linear stage of softening with a low negative coefficient θ.

Comparison of the flow curves of polycrystalline nanostructured titanium shows that at deformation temperatures below 623 K, the deformation behaviour of these materials greatly differs both qualitatively (in the stages) and quantitatively (by the value of the yield limit, the strain hardening coefficient, plasticity, the magnitude of applied stresses). At a temperature of 623 K, the flow curves of nanostructured and recrystallised titanium do not differ qualitatively and are characterised only by different duration of each stage of plastic deformation and, correspondingly, by different total plasticity. Consequently, starting at a temperature of 623 K, the internal structure in nanostructured titanium produced by special treatment changes so that its deformation behaviour becomes approximately identical with that of recrystallised titanium.

The variation of the form of the flow curves of nanostructured titanium with the increase of deformation temperature from room temperature to 623 K, and also in comparison with polycrystalline titanium, is in good agreement with the behaviour of the quantity $\sigma_{0.2}/\sigma_B$ (where $\sigma_{0.2}$ and σ_B is the yield limit and maximum stress, respectively). In [239], this criterion is regarded as the criterion of resistance to homogeneous plastic deformation on the macroscale level: at $\sigma_{0.2}/\sigma_B \rightarrow 1$, the susceptibility to the development of localised plastic deformation on the macrolevel increases.

The temperature dependence of $\sigma_{0.2}/\sigma_B$ for nanostructured and recrystallised titanium, shown in Fig. 4.36, shows that the margin

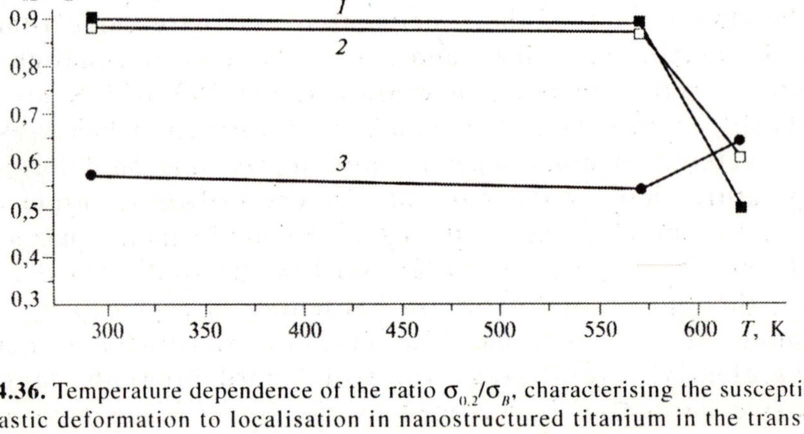

Fig. 4.36. Temperature dependence of the ratio $\sigma_{0.2}/\sigma_B$, characterising the susceptibility of plastic deformation to localisation in nanostructured titanium in the transverse (*1*) and longitudinal (*2*) directions in relation to the pressing axis, and in coarse-grained titanium (*3*).

of resistance to homogeneous plastic deformation on the macrolevel in nanostructured titanium at temperatures from 293 to 573 K inclusive is very low $\sigma_{0.2}/\sigma_B$ = 0.97–0.85

The variation of the form of the yield curves of nanostructured titanium with an increase of deformation temperature from room to 623 K and also in comparison with polycrystalline titanium is in good agreement with the behaviour of $\sigma_{0.2}/\sigma_B$ (where $\sigma_{0.2}/\sigma_B$ is respectively the yield limit and maximum stress). In [239], these quantities regarded as a criterion of the resistance of the homogeneous plastic deformation on the macroscale level: at $\sigma_{0.2}/\sigma_B \to 1$, the susceptibility to the development of localised plastic deformation on the macrolevel increases. At the same time, in recrystallised titanium $\sigma_{0.2}/\sigma_B$ at these temperatures does not exceed 0.6. At T = 623 K, the $\sigma_{0.2}/\sigma_B$ ratio of nanostructured titanium decreases to the value of this parameter for recrystallised titanium. On the basis of these data it is proposed in [239] that, starting at a temperature of 623 K, nanostructured titanium should show a large decrease of the susceptibility to the localisation of plastic deformation on the macrolevel because the main features of the development of the strain relief should become the same as in recrystallised titanium.

It was reported in [198] that the characteristic feature of deformation in the tensile loading of nanostructured titanium, produced by equal-channel angular pressing, is the development of localised deformation bands. In tensile loading at a rate of $3.3 \cdot 10^{-3}$ s^{-1}, the start of plastic deformation of the specimens, cut in the direction normal to the direction of equal-channel angular pressing, the surface shows the formation of bands of localised deformation several micrometres wide propagating under an angle close to 60° in relation to the tensile loading axis. These bands propagate almost instantaneously through the entire cross-section of the specimen. When reaching of the tensile strength on the surface of the specimen, two macrobands of localised deformation approximately 0.5 mm wide form on the surface of the specimen in the gauge part. The bands are located under the angle of ~120° in relation to each other and under the angle of approximately 60° in relation to the tensile loading axis. The specimens failed by shearing along one of the macrobands of localised deformation. In plastic deformation of the titanium specimens cut in the direction parallel to the direction of equal-channel angular pressing, the formation of macrobands of localised deformation is preceded by the formation of a neck. On the flow curve, as in the case of the specimens, cut in the direction normal to the direction of equal-channel angular pressing, the formation of macrobands of localised deformation on the flow curve corresponds to the start of the stage of decreasing stress. The formation of the neck on the surface of the specimens is accompanied by the formation of mesobands of localised deformation. However, in the investigated case, they develop gradually, are discontinuous and situated under a small angle (0–15°) to the direction of tensile loading. The meso- and macrobands of localised deformation developed during tensile loading also in the specimens after preliminary (prior to deformation) pre-recrystallisation annealing in the temperature range 473–673 K. The meso- and macrobands of localised deformation do not form during tensile loading on the surface of the recrystallised titanium specimens. It should be mentioned that the authors have noted the formation of macrobands of localised deformation also in nanostructured nickel produced by equal-channel angular pressing.

Similar investigations of the evolution of the deformation relief were carried out in the tensile test at a rate of $3\cdot10^{-5}$ s^{-1}. At temperatures of increased susceptibility of nanostructured titanium to the localisation of plastic deformation on the macrolevel there is agreement between the stages of the flow curve and the nature of localisation of plastic deformation on the meso- and macrolevel. This is especially clear at T = 573 K. At this temperature, the strain relief in the form of folds (microbands of localised plastic deformation) 10–20 μm long forms at the degree of plastic deformation of approximately 2%, i.e. at the start of the linear stage on the flow curve. These microbands (folds) are oriented under angles close to 60° in relation to the direction of tensile loading. In subsequent stages, the length of the microbands increases and in certain localised areas the density of the microbands also increases. One of these areas becomes the mesoband of localised plastic deformation embracing a large part of the cross-section of the specimen. Inside the mesoband the formation of long microbands of localised plastic deformation continues. The orientation of these microbands along the entire length becomes either constant or changes many times. In the final analysis, the increase of the degree of deformation increases the length and width of the mesoband of localised plastic deformation, and at $\varepsilon \approx$ 8–9%, the mesoband intersects the entire cross-section of the specimen resulting in the formation of a macroband of localised plastic deformation (Fig. 4.37*a*). On the flow curve, this corresponds to the moment of completion of the linear stage of

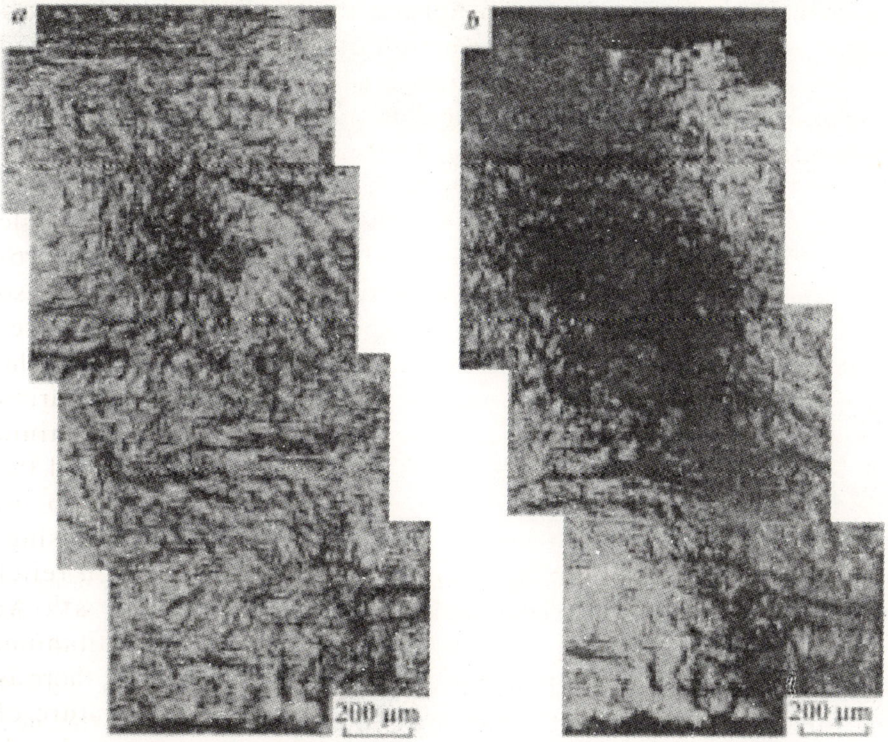

Fig. 4.37. Formation of a macroband of localisation of plastic deformation in nanostructured titanium deformed at T = 573 K. *a*) = 9, *b*) = 14.5%.

strain hardening and the start of the linear stage of strain softening (Fig. 4.34, 4.35). In the areas of exit of the investigated macroband of the side surfaces of the specimen there is a small decrease of the width of the symmetric form, i.e. necking starts. With further deformation, the density of the mesobands inside the macro band increases and at ε = 14–15% in some areas of the macroband the internal structure of the latter is no longer visible at the magnification used in the investigations (Fig. 4.37*b*). In this case, the width of the macroband and its orientation in relation to the direction of tensile loading are almost constant (equal to approximately 500–600 μm and 60°, respectively).

The stage of development of the macroband of localised plastic deformation with the formation of a neck on the side surfaces of the specimen corresponds to the linear section of strain softening (Fig. 4.35*b*). In this stage, the necking remains symmetric with a small depression; at a width of the working part of the specimen of 2500 μm, the depression is not greater than

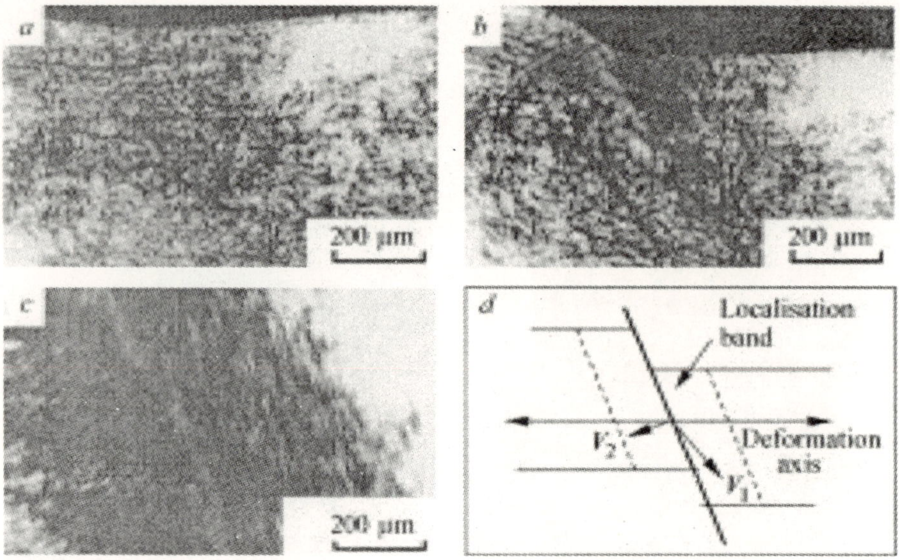

Fig. 4.38. Symmetric (*a*) and non-symmetric (*b*) necking, and also shearing of parts of the specimen in relation to each other (*a*), and the shearing diagram (*d*); T = 573 K.

100 μm. Further localisation of plastic deformation on the macrolevel does not take place by simple narrowing in the neck with subsequent rapture by the formation of a crack; it takes place by 'shifting' of one part of the specimen in relation to the other along the central section of the macroband (Fig. 4.38). Consequently, necking becomes asymmetric and this is manifested on the flow curve by the formation of another linear stage of strain hardening. The directions of movement of the material to different sides from the 'shearing' surface differ (Fig. 4.38*c*). The scheme of 'shearing' is shown in Fig. 4.38*d*. On one side from the 'shearing' surface, the direction of movement of the material forms a large angle with the direction of tensile loading, and on the other side the direction of movement is close to the direction of tensile loading and the speed of movement is approximately 50% lower. In the immediate vicinity of the 'shearing' surface, the movement of the material is vortex-like, with the direction of movement changing from V_1 to V_2. During deformation, one of the directions of movement is maintained and the other one gradually approaches the direction of tensile loading. As a result of this process, the specimen is separated into parts without the formation and opening of the crack. This failure is very similar to that detected in single crystals of aluminium and copper in which the process of deformation at 4.2 K with a slightly developed neck is accompanied

by the formation of narrow zone is of intensive shearing in the primary or secondary slip system, where deformation was localised until it was concentrated in a single plane. Fracture took place after marked elongation with the formation of a slightly developed neck [242].

At room temperature, the localisation of plastic deformation in nanostructured titanium developed in the same manner as at a temperature of 573 K (Fig. 4.39).

As shown in Fig. 4.36, the susceptibility to localisation of plastic

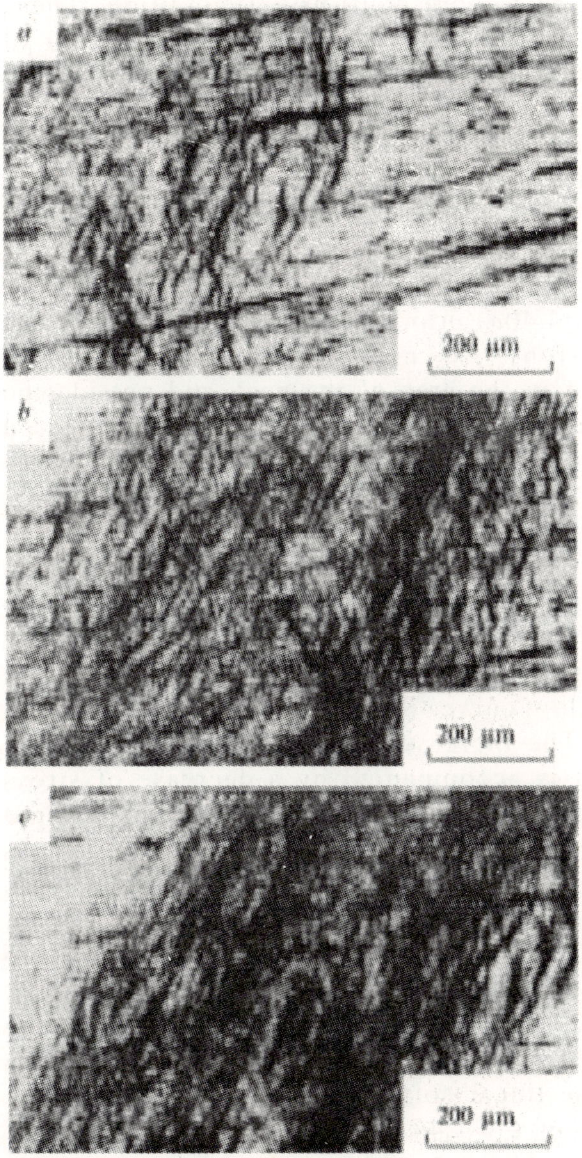

Fig. 4.39. Bands of localisation of plastic deformation in nanostructured titanium at 293 K. *a*) 0.5, *b*) 1, *c*) 2.5%.

deformation greatly decreases when the deformation temperature is increased to 623 K. According to all features, in addition to the greatly increased plasticity, the flow curves of nanostructured titanium at a temperature of 623 K are identical with those of recrystallised titanium. According to the results published in [150], the structure of nanostructured titanium after severe plastic torsional deformation at a temperature of 623 K is characterised by qualitative changes: the diffusion activity of the grain boundaries increases and this is accompanied by grain boundary recovery, i.e. grain boundaries change to a more equilibrium condition. These processes may result in changes of the deformation behaviour of nanostructured titanium and the properties of this material become similar to those of the recrystallised material. According to the data in [198], extensive structural changes and grain growth to 2–4 μm are observed after isothermal annealing at 773 K, but in the conditions of active plastic deformation these processes may start at lower temperatures.

No features of meso- and macrolocalisation of plastic deformation were observed (Fig. 4.40*a*) prior to necking (a wide meniscus with a small depression) in nanostructured titanium at 673 K. The surface relief in the area of the neck and outside the area does not differ. With further deformation a crack appeared either in the centre of the neck (Fig. 4.40*a*) or at one of the edges of the neck (Fig. 4.40*c*, *d*) and, growing in steps and expanding, resulted in failure of the specimen.

The development of the deformation relief in coarse-grained titanium took place by the following mechanism. At room temperature in deformation there was no meso- and macrolocalisation of deformation up to $\varepsilon = 20\%$. This corresponds to the maximum stress on the flow curve. Small narrowing was detected at a strain of $\varepsilon = 20\%$. The increase of the degree of narrowing with further deformation is accompanied by a decrease of stress up to the moment of the formation of a crack with subsequent rapid failure of the specimen (Fig. 4.41*a*, *b*).

At a temperature of 573 K no features of localisation were detected up to the maximum on the flow curve ($\varepsilon = 14\%$). Narrowing was detected at $\varepsilon = 14\%$ and this area showed a dip at $\varepsilon = 16\%$ with subsequent failure through this region.

At a temperature of 623 K at $\varepsilon = 9\%$, the specimen showed the formation of a small meniscus (up to 2 mm wide) and the flow curve showed in this case linear softening with a low coefficient ($\cong$150 MPa). At a strain of $\varepsilon = 24\%$ a crack appeared at one of

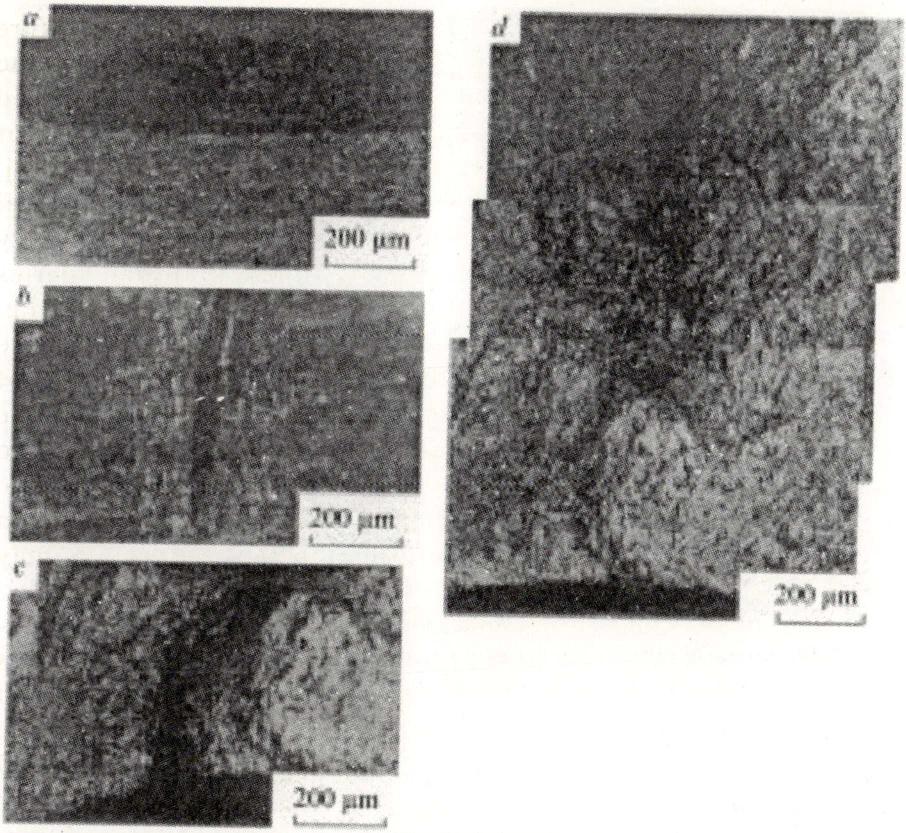

Fig. 4.40. Surface relief of nanostructured titanium, deformed at T = 623 K (*a–d*) and formation of a localisation band ahead of the crack at T = 633(*c*). *a*) = 22, *b*) 42, *c*) 50, *d*) 40%.

the edges of the specimen and propagated to a length of up to 400 μm and was blunted to approximately 300 μm in the diameter, forming ahead in the remaining cross-section a band 400–600 μm wide characterised by displacement of one part of the specimen in relation to another part (Fig. 4.41*c, d*). The band showed no special contrast nor localisation mesobands. The structure did not differ from the structure of the remaining part of the specimen. The band was detected only on the basis of the movement of the bright areas of the surface of the specimen, recorded on the photographs. The formation of this band is accompanied by a large decrease of the stress and subsequent failure of the remaining part of the specimen without crack opening (Fig. 4.41*d*). The same type of failure was observed in one of the specimens of nanostructured titanium at a deformation temperature of 623 K. This is in good agreement with the data presented in Fig. 4.36 where the values of $\sigma_{0.2}/\sigma_B$ for the

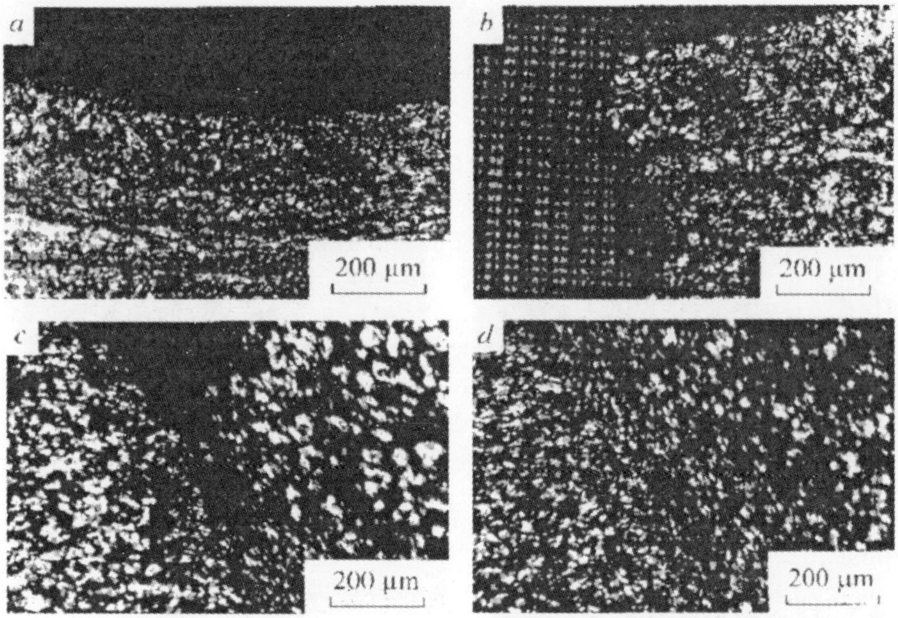

Fig. 4.41. Surface relief of coarse-grained titanium deformed at room temperature (*a*, *b*) and at 623 K (*c*, *d*). *a*) = 24%, *b*) after fracture, *c*) = 29%, *d*) 'shearing'.

nanostructured and recrystallised titanium and the same temperature are identical.

Thus, nanostructured titanium is characterised by high susceptibility to localisation of plastic deformation in the entire range of thermal stability of its structure (293–573 K). At a temperature of 623 K when the nanostructure becomes unstable, the mechanical properties of nanostructured and recrystallised titanium become similar, like the development of the strain relief. The formation of the macroband of localised plastic deformation in nanostructured titanium corresponds, on the flow curve, to the linear stage with a low strain hardening coefficient. The completion of the stage corresponds to the exit of the band to the side surfaces of the specimens and to the appearance of symmetric necking in this area. The change of the form of necking from symmetric to non-symmetric takes place during the displacement of two parts of the specimen in relation to each other along the axis of the strain macroband and is accompanied by further failure without crack opening. The strain behaviour and localisation of plastic deformation in the specimens of nanostructured titanium, cut along and across the pressing direction, differ only quantitatively. No qualitative difference was found in the development of these processes. From the moment of the start of development of the macroband of localised plastic deformation, the formation of the individual stages of the flow curve of nanostructured titanium reflects the development of plastic deformation in this band.

The meso- and macrobands of localised shearing were also found previously in deformation of coarse-grained [243] and nanostructured materials [239, 244, 245]. According to the authors who examined this phenomenon, the reasons for this phenomenon differ. For example, Kaibyshev [243] links the formation of the mesobands of localised shearing with the presence of a texture, assuming that the tilt boundaries in the textured material are a less efficient obstacle to the movement of this location is in comparison with the normal high-angle boundaries. The authors of [244] examined the formation of macrobands of localised shearing in deformation of nanostructured materials, produced by gas phase deposition, and assumed that the formation of these bands is caused, as in metallic glasses, by the presence of pores and non-equilibrium vacancies, as indicated by the increased specific volume of this material in comparison with the appropriate tabulated values.

The method of production of the nanostructured materials by the method of equal-channel angular pressing prevents the formation of pores in the material which could increase the specific volume. However, dilatometric investigation show that the nanostructured materials subjected to the effect of severe plastic deformation, like the nanostructured materials produced by gas phase deposition or by compacting disperse powders, are characterised by a larger specific volume in comparison with the coarse-grained materials. This may be caused by the greater length of the grain boundaries and by the defective structure of the grain boundaries in these materials [237]. Consequently, the formation during tensile loading of meso- and macrobands of localised shearing on the surface of the specimens of the nanostructured materials after the effect of severe plastic deformation may be associated with the presence of defects resulting in a larger specific volume. On the other hand, it is well known [202] that severe plastic deformation forms nanostructures with a distinctive texture which, according to [244], maybe the reason for the formation of meso- and macrobands of localised shearing in deformation.

Composites

Recently, extensive investigations have been carried out to examine the mechanical properties in tensile and compressive loading of composites produced by the consolidation of a mixture of fine-dispersion powders. The experimental results [221], obtained in compression of Cu–3 vol% TiC and $CuTi_4$–10 vol% TiC composites, show that the presence of hardening nanosized particles of TiC increases the yield limit ($\sigma_{0.2}$) of Cu and Cu TiC_4 1.5–2 times whilst retaining high plasticity (>10%). At the same time, the results of room temperature tensile tests show [187] that, irrespective of preliminary treatment (Table 4.7), failure of Cu–3vol% TiC composite is brittle, on the elastic part of the stress–strain (flow) curve. The scatter in the values of tensile strength (σ_B) reaches 100%. However, in all investigated conditions, the σ_B value of the Cu–3 vol% TiC composite is higher in comparison with pure copper, in both the coarse-grained and nanostructured condition (see Table 4.7).

The large increase of the strength properties of copper in compression is also found when introducing the nanosized particles of Al_2O_3 (~14 nm) [224]. The authors of [224] reported that the higher values of $\sigma_{0.2}$ for the Cu–3 vol% Al_2O_3 composite in comparison with pure copper are detected not only at room but also higher test temperatures (Table 4.8). At the same time, the introduction into the copper matrix of the Al_2O_3 particles with the mean size of approximately 1 μm does not increase the values of $\sigma_{0.2}$ [224].

The comparison of the evolution of the structure of these composites during annealing in the tempe-rature range 473–1273 K

Table 4.7. Mechanical properties of copper and ultrafine-grained Cu–3 vol% Ti composites at room temperature

Material	σ_B, MPa	ε, %
Pure annealed copper	232	~50
Pure nanostructured copper	369	6.2
Cu–3 vol.% TiC (extrusion)	350–300	0
Cu–3 vol.% TiC (extrusion + torsional deformation)	640–400	0
Cu–3 vol.% TiC (extrusion + torsional deformation + annealing for 1 h at 673 K)	605–401	0
Cu–3 vol.% TiC (extrusion + torsional deformation + annealing for 1 h at 873 K)	620–350	0

Table 4.8. Yield limit of copper and Cu–Al_2O_3 composite after hot extrusion at different test temperatures [224], MPa

Material	Temperature, K			
	298	713	763	813
Pure copper	107	~45	–	–
Cu with Al_2O_3 micron-sized particles	103	~45	–	–
Cu–1 vol.%Al_2O_3	125	87	77	67
Cu–3 vol.%Al_2O_3	165	113	90	81

and of the strength properties enabled the authors of [224] to conclude that the values of $\sigma_{0.2}$ of the ultrafine-grained composites depend not only on the particle size of the hardening phase but also on the particle size of the metallic matrix.

The dependence of $\sigma_{0.2}$ of ultrafine-grained composites, produced by the consolidation of the mixture of powders and subsequent hot extrusion, on the mean grain size of the metallic matrix, was examined in greater detail in [223] on the example of Ag–5 vol% Cr_2O_3 composite. The authors of [223] found that the dependence of $\sigma_{0.2}$ of this composite on the mean grain size of silver (the value of d was varied in the range from 0.02 to 8 μm) is efficiently described by the Hall–Petch law (Fig. 4.42). In this case, the Hall–Petch coefficient, determined on the basis of the angle of inclination of the curve of the dependence of $\sigma_{0.2}$ on $d^{-1/2}$ (Fig. 4.42) was equal to 0.56 MPa·cm$^{-1/2}$, which corresponds to the Hall–Petch coefficient for pure silver.

A number of investigations of the mechanical properties was carried out on the composites produced by internal oxidation and consolidation of a mixture of fine-grained powders followed by hot extrusion [228–229]. The experiments were carried out on the specimens of Cu–0.5 wt% Al_2O_3 composite in three conditions: immediately after extrusion (condition 1), after extrusion and subsequent cold deformation by rolling to 70% (condition 2) and extrusion and subsequent severe torsional plastic deformation by the pressure + shear method (the true logarithmic strain ~7) (condition 3 – the nanostructured composite).

The tensile test at room temperature carried out on the Cu–0.5 wt% Al_2O_3 composite in condition 1 showed no anomalous form of the flow curves (Fig. 4.43, *1*). After the effect of severe

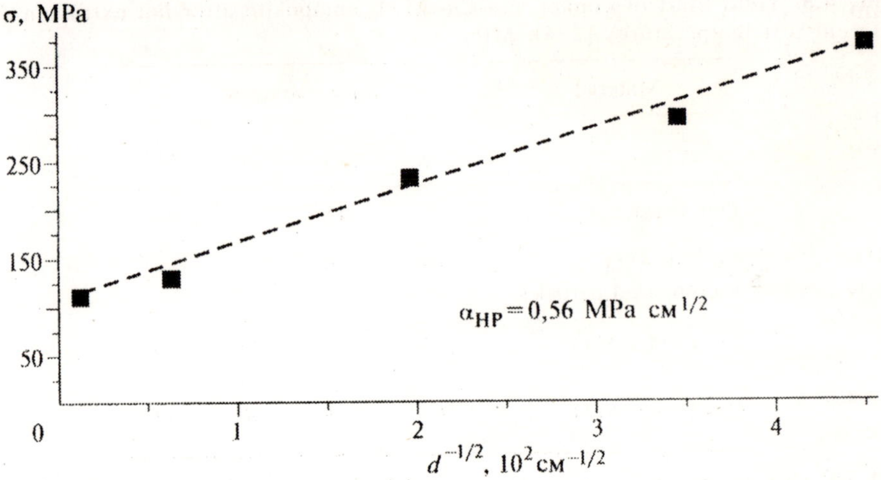

Fig. 4.42. Dependence of the yield limit of Ag–5vol% Cr_2O_3 composite at room temperature on the grain size [223].

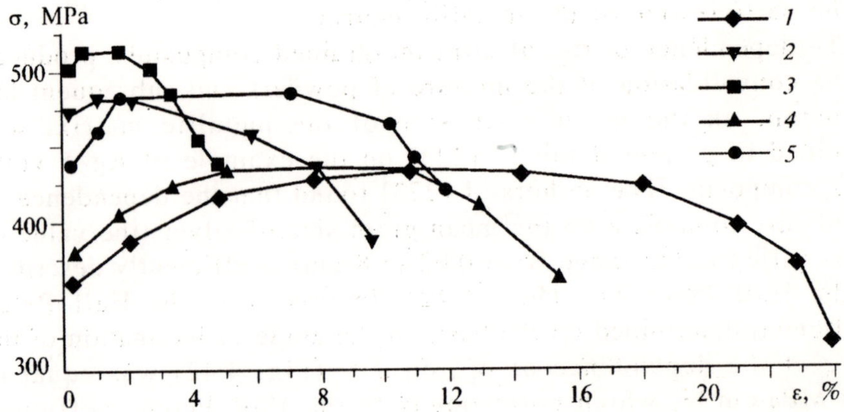

Fig. 4.43. The flow curves of the specimens of Cu–0.5 wt% Al_2O_3 composite at room temperature. 1) condition 1, 2) 2, 3) 3, 4) condition 2+ annealing at T = 673 K, 5) condition 3+ annealing at T = 673 K.

plastic torsional deformation, the Cu–0.5 wt% Al_2O_3 nanostructured composite, like nanostructured copper [246], showed no extensive strain hardening (Fig. 4.43, *3*). The small strain hardening was also detected in the tensile loading of the Cu–0.5 wt% Al_2O_3 composite (Fig. 4.43, *2*). Pre-recrystallisation annealing in the temperature range 423–673 K did not change the form of the stress–strain (flow) curve of the Cu–0.5 wt% Al_2O_3 composite in condition 1 but resulted in the effect of strain hardening on the stress–strain curves of the investigated composite in the conditions 2 and 3 (see Fig. 4.43, *2* and *3*).

Examination of the kinetics of development of plastic deformation in the tensile loading of the Cu–0.5 wt% Al_2O_3 composite shows that its nature depends greatly on the existing structure. In the Cu–0.5 wt% Al_2O_3 composite in condition 1, a homogeneous flow is detected from the very start of plastic deformation. This flow is accompanied by the formation of a small neck. When reaching the tensile strength on the surface of the composites, as in pure nanostructured materials, two macrobands of localised deformation gradually developed under the angle of ~60° in relation to the direction of applied loading. The specimen failed by shearing along one of the macrobands of localised deformation. The macrobands of localised deformation in tensile loading also developed on the surface of the specimens of the Cu–0.5 wt% Al_2O_3 composite in the nanostructured condition. In this case, the macrobands of localised deformation formed from the very start of plastic flow, and almost the entire deformation fracture was localised in a narrow shear band. The nanostructured composite also failed by shearing along one of the bands of localised deformation. After pre-recrystallisation annealing at a temperature of 423–673 K, the kinetics of development of plastic deformation of the Cu–0.5 wt% Al_2O_3 nanostructured composite is identical with that of the composite in condition 1.

A different mechanism of development of plastic flow is observed in the Cu–0.5 wt% Al_2O_3 composite in condition 2. After deformation by rolling to 70%, the surface of the specimens of the Cu–0.5 wt% Al_2O_3 composite did not show the formation of macrobands of localised deformation. From the very start of plastic deformation, there was a homogeneous flow with subsequent formation of a small neck consisting of mesobands of localised deformation, spreading throughout the entire specimen in the direction normal to the tensile loading axis. The start of necking coincides with the start of the stage of decreasing stress on the stress–strain curve. Failure of the Cu–0.5 wt% Al_2O_3 composite in condition 2 takes place by separation on the plane, normal to the tensile loading axis.

Thus, the nanostructured composites, like the pure nanostructured materials, are characterised by the development of the bands of localised deformation.

The tensile tests results show (Table 4.9) that the formation of the nanostructure in the Cu–0.5 wt% Al_2O_3 composite under the effect of severe plastic torsional deformation by torsion increases the tensile strength (σ_B) and yield strength ($\sigma_{0.2}$) of the composite

Table 4.9. Mechanical properties of on temperature of pure copper and Cu–0.5 wt% Al_2O_3 composite

Material (condition)	σ_B	$\sigma_{0.2}$	Δ, %
	MPa		
Cu – 0.5 wt.% Al_2O_3 (1)	422–430	353–360	24.8
Cu – 0.5 wt.% Al_2O_3 (2)	487	475	7.2
Cu – 0.5 wt.% Al_2O_3 (3)	523–499	489–463	5.9
Nanostructured copper	367	352	6.2
Recrystallised copper	255–230	175–190	50

at room temperature by 20 and 35%, respectively. In this case, the strain to fracture decreased by up to four times. Deformation by rolling to 70% also increased σ_B and $\sigma_{0.2}$. However, in the latter case, $\sigma_{0.2}$ increased by a smaller value in comparison with the effect of severe plastic torsional deformation in torsion (see Table 4.9). The annealing of the Cu–0.5 wt% Al_2O_3 nanostructured composite at temperatures of 673 and 873 K for 1 h reduces σ_B and $\sigma_{0.2}$ and increases (Fig. 4.44*a*, *b*). Even after annealing at T = 873 K, the strength properties of the nanostructured Cu–0.5 wt% Al_2O_3 composite were twice as high as in the recrystallised copper and remain on the level of the strength properties of the Cu–0.5 wt% Al_2O_3 composite in condition 1.

The experimental results obtained for the yield limit in the tensile loading of Cu–0.5 wt% Al_2O_3 composite and pure copper were used for the analysis of the nature of hardening of the Cu–0.5 wt% Al_2O_3 composite [247]. Theoretical calculations of the $\sigma_{0.2}$ values of the given composite were carried out using Orowan's relationship for dispersion hardening [248]:

$$\sigma_{0.2} = \sigma^{m}_{0.2} + (45Gb/2r)\cdot f,$$

where $\sigma^{m}_{0.2}$ is the yield limit of the matrix without the particles; b is the Burgers vector of the dislocation in copper ($b = 256\cdot10^{-10}$ m); f is the volume fraction of the particles of Cu–0.5 wt% Al_2O_3 (f = 0.74 without taking into account the large precipitates of Al_2O_3 which do not contribute to hardening); r is the radius of the Al_2O_3 particles (20 nm); G is the shear modulus (for a copper matrix

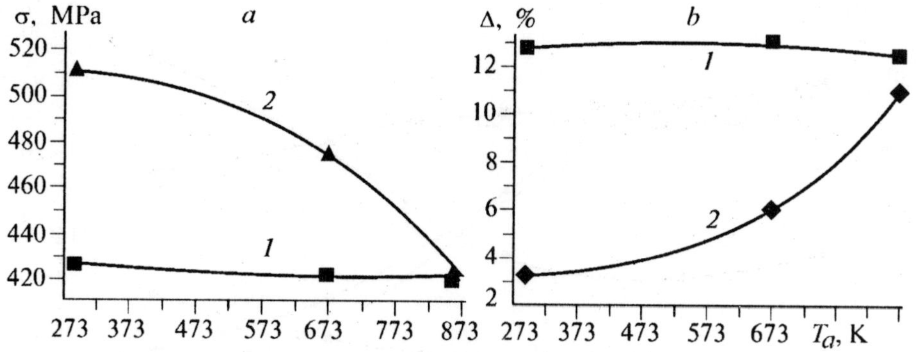

Fig. 4.44. Dependence of tensile strength (*a*) and deformation to fracture (*b*) of the Cu–0.5 wt% Al_2O_3 composite in the condition (*1*) and in nanostructured condition (*2*) on preliminary annealing temperature.

Table 4.10. Experimental and calculated (using the Orowan model) values of the yield limit of Cu–0.5 wt% Al_2O_3 composite.

Material (condition)	$\sigma_{0.2}$, MPa			Difference, %
	Experimental	Experimental marix	According to Orlov	
Cu – 10 wt.% Al_2O (1)	360	150	362	8
Cu – 10 wt.% Al_2O_3 (2)	475	384	563	15
Cu – 1wt.% Al_2O_3 (3)	489	351	524	10

$4.2 \cdot 10^4$ MPa). Table 4.10 shows that the calculated values of $\sigma_{0.2}$ are in good agreement with the experimental values.

Examination of the temperature dependence of the mechanical properties of ultrafine-disperse composites was carried out on the example of Cu–0.5 wt% Al_2O_3 composite in condition 2. Figure 4.45 shows the temperature dependence of tensile strength, yield strength and strain to fracture of the composite (curves 1'–3') and pure nanostructured copper (curves 1–3) produced by the effect of severe plastic deformation using the method of equal-channel angular pressing. It may be seen that at T = 293 and 473 K, the values of the tensile and yield strength of the composite in condition 2 were close to the values for nanostructured copper and 2 and 6 times higher than the values for coarse-grained copper. With the increase of temperature, the values of σ_B and $\sigma_{0.2}$ for the composite and pure nanostructured copper monotonically decrease but the thermal stability of the mechanical properties of the

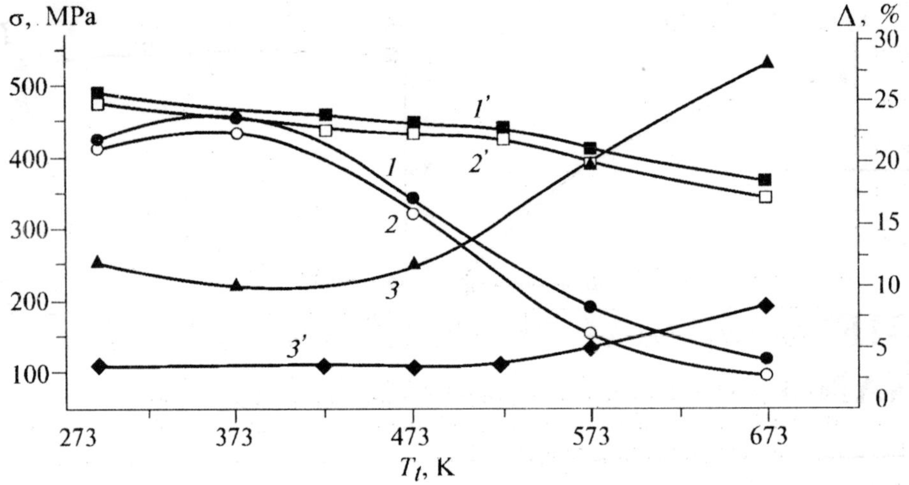

Fig. 4.45. The dependence of ultimate tensile strength (*1*, *1'*), yield strength (*2*, *2'*) and plasticity (*3*, *3'*) on test temperature for nanostructured copper (*1–3*) and Cu–0.5 wt% Al_2O_3 composite (1'–3').

composite is considerably higher than that of the nanostructured copper. This is associated evidently with the presence of dispersed Al_2O_3 particles in the volume of the composite. The particles may block movement of the dislocations and migration of the grain boundaries, preventing recrystallisation, leading to softening.

Thus, the examined results of experimental investigations of the mechanical properties of nanostructured metals and ultrafine-dispersed composites with a metallic base make it possible to draw the following conclusions.

The nanostructured metals and nanostructured composite based on them, produced by the effect of severe plastic deformation, have improved strength characteristics in comparison with the appropriate coarse-grained and ultrafine-grained materials. The strength properties in tensile and compressive loading of the nanostructured properties are 2–3 times higher in comparison with the properties of the pure nanostructured metals.

Strain behaviour in the yield limit of the nanostructured metals, produced by severe plastic deformation, depend strongly on the state of the grain boundaries at the same grain size, indicating that it is necessary to take into account the role of the defect structure of the grain boundaries when analysing fulfilment of the Hall–Petch relationship in these materials.

The development of the bands of localised deformation is the characteristic mechanism of deformation at room temperature of nanostructured materials produced by severe plastic deformation.

Chapter 5

DIFFUSION-ACTIVATED CREEP OF NANOSTRUCTURED MATERIALS

The prospects for using nanostructured materials produced by severe plastic deformation in industry determines the importance of examining special features of the creep of these materials at elevated temperatures. Such investigations have been carried out only in a small number of studies [249–253]. A considerably larger number of studies have been concerned with examining the relationship of the creep of nanostructured materials produced by electrodeposition or powder consolidation [225, 244, 245–260]. However, the results are contradicting.

We discuss the literature and original results of the examination of the relationships governing the creep of nanostructured materials, including those produced by severe plastic deformation. Special attention is given to examining the effect of the condition of the grain boundaries on the creep relationships.

5.1. RELATIONSHIPS GOVERNING THE CREEP OF NANOSTRUCTURED MATERIALS

Pure metals

In the first study [260] concerned with the examination of the relationships governing the creep of electrodeosited nanostructured copper (d ~30 nm) it was established that at room temperature this process is described by the Coble law of diffusional creep [261].

A more detailed examination of creep in the temperature range (0.22–0.24) T_m of nanostructured copper in the same structural state as described previously, was carried out in [225]. In addition to the observed linear dependence of the creep rate in the steady

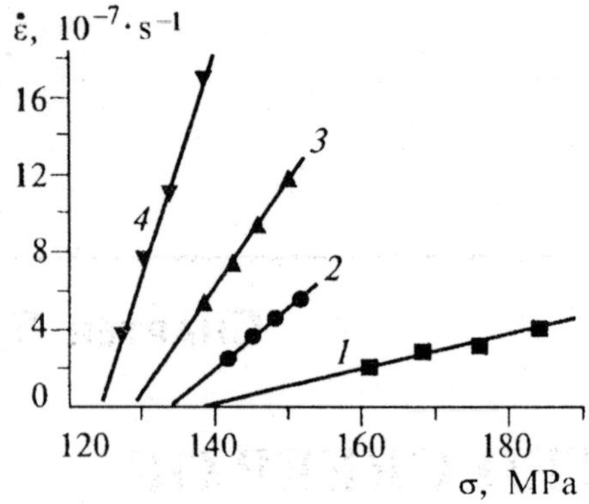

Fig. 5.1. Dependence of the steady-state creep rate on applied stress for nanostructured copper [225].
1) T = 293 K, *2*) 303, *3*) 313, *4*) 323 K.

stage on stress, some other special features of creep in comparison with coarse-grained nickel were detected. Firstly, for every test temperature, examination showed the existence of a threshold stress which decreased with increasing temperature (Fig. 5.1). Secondly, the measured activation energy of creep of nanostructured copper (70 kJ/mole) was considerably lower than the activation energy of volume (196 kJ/mole) and grain boundary (104 kJ/mole) self-diffusion in copper and was approximately identical with the activation energy of diffusion at the grain boundaries in nanostructured copper (60 kJ/mole) [262]).

On the basis of the experimental results, the authors of [225] assumed that the creep of nanostructured copper is controlled by grain boundary diffusion whose rate is in turn controlled by emission or adsorption of point defects by the grain boundaries. In this case, the presence of the threshold stress in creep may be associated with the higher fraction of equilibrium grain boundaries in the nanostructured material which, according to Gleiter et al [262], cannot operate as effective sources and sinks of point defects, if the energy is lower than some value. Consequently, the threshold stress is the stress required for a local increase of the energy of the grain boundaries [225]. In accordance with these results, the dependence of the creep rate in the steady stage on stress for the nanostructured materials, according to the authors of [225] is described by the modified Coble equation:

$$\dot{\varepsilon} = \frac{148\,D_0\,\delta\Omega\sigma_{et}}{\pi\,d^3kT}\exp(Q_b^*/kT), \tag{5.1}$$

where D_0 is the pre-exponential multiplier in the equation for the coefficient of grain boundary diffusion; δ is the width of the boundary; Ω is the atomic volume; σ_{ef} = σ_0 + (σ_0 the threshold stress, σ is applied stress); d is the grain size; Q^*_b is the activation energy of grain boundary diffusion for of the nanostructured condition; R and T have the generally recognised meaning.

A different conclusion was obtained by Mishra [251] when comparing the experimental dependence $\dot{\varepsilon}(\sigma)$ for nanostructured copper produced by electrodeposition with the dependence predicted by the theory of Coble diffusional creep [261]. Mishra shows that the experimentally determined value of the exponent at the stress index (n) for nanostructured copper is higher ($n > 4$) than predicted by the theory of Coble diffusional creep [261] ($n = 1$). It was established that the experimentally determined creep rates of nanostructured copper in the steady stage at the given stresses are considerably higher than those indicated by calculations carried out using the theory proposed in [261] and the concept of threshold stress [225]. Identical results were obtained in [253] also for nanostructured copper formed by the effect of severe plastic deformation. However, in [253] in the examination of the creep of nanostructured iron, produced by the effect of severe plastic deformation, it was established that the experimentally recorded creep rates in the steady stage are however lower than those indicated by the theory of Coble diffusional creep [261]. The experimental value of n, as in the case of nanostructured copper, was higher than the theoretical value for Coble creep.

The attempt to generalise the literature experimental data on the dependence of the steady state creep rate on stress for the nanostructured materials, produced by different methods, and compare these results with the values calculated from the theory of Coble diffusional creep and grain boundary sliding, controlled by grain boundary diffusion, was made in [251]. The author of [251] used the following equation for the calculation of the rate of diffusional creep [261]:

$$\dot{\varepsilon} = 47.7\frac{\Omega\delta D_b\sigma}{kTd^3}, \tag{5.2}$$

Here D_b is the coefficient of grain boundary diffusion; the meaning of the remaining notations is the same as in equation (5.1).

The theoretical values of the creep rate, controlled by grain boundary sliding, were calculated from the the model proposed in [263):

$$\dot{\varepsilon}_{\text{GBS}} = 2\cdot 10^5\frac{D_b b^4\sigma^2}{GkTd^3}, \tag{5.3}$$

Where G is the shear modulus; b is the Burgers vector.

When comparing the equations (5.2) and (5.3) with the experimental relationships $\dot{\varepsilon}(\sigma)$ for different nanostructured materials (Figs. 5.2, 5.3), Mishra [251) concluded that neither of the two investigated theories correspond to the experimental data. In particular, the experimentally recorded values of the exponent at the

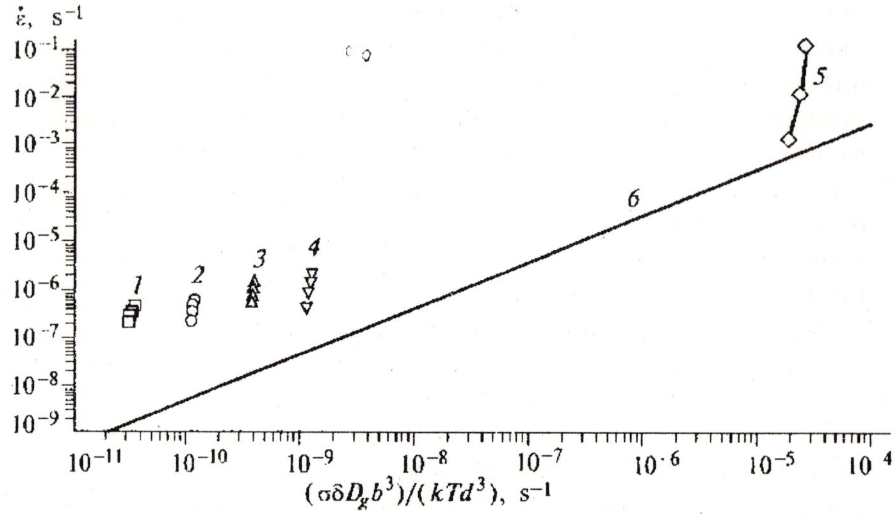

Fig. 5.2. Comparison of experimental data for pure nanostructured copper with the theoretical rate of diffusional creep [252]. *1*) T = 293 K, *2*) 303, *3*) 313, *4*) 323, *5*) 523 K; *6*) the theoretical Coble creep curve.

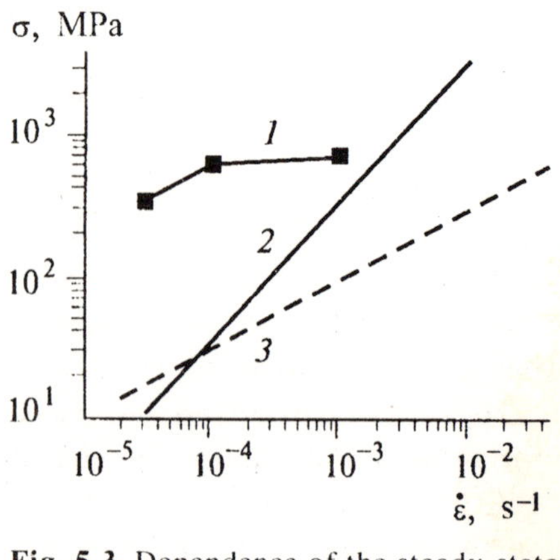

Fig. 5.3. Dependence of the steady-state creep stress for nanostructured iron [252]. *1*) experimental data for nanostructured iron; *2*) theoretical Coble creep curve; *3*) theoretical curve of grain boundary sliding.

stress index (n > 4) for all nanostructured materials, investigated in the study, regardless of the method of production, are considerably higher than the theoretical values (n = 1 [261], n = 2 [263]). The experimentally recorded steady state creep rates for specific nanostructured materials (for example, nanostructured copper) were higher than for others (nanostructured iron) - lower than the values corresponding to the equations (5.2) and (5.3), regardless of the fact that these nanostructured materials were produced by the same method (by severe plastic deformation) and the grain size was 100 and 150 nm, respectively. According to the author of [251], these contradictions in the behaviour of nanostructured copper and iron in creep are caused by the different temperatures and stresses used in the experiments.

It has already been mentioned that as a result of the special features of formation, the nanostructures, produced by severe plastic deformation are characterised not only by a longer length but also by distinctive non-equilibrium nature of the grain boundaries. Pre-recrystallisation annealing, and sometimes simply long-term holding at room temperature, results in the recovery of the defective structure of the grain boundaries and in the transition of the grain boundaries to a more equilibrium condition. Usually [264, 265], this transition results in a change of the kinetic parameters of certain processes such as grain boundary mass transfer and grain boundary sliding. These processes may influence creep. It is natural to assume that the differences in the investigated relationships of creep of nanostructured copper and iron, produced by the effect of severe plastic deformation, and noted by the authors in [251, 253], are associated with the different condition of the grain boundaries in the materials investigated in different studies. This assumption is supported by the data obtained by the authors in examining the nature of creep of nanostructured copper and nickel, produced by the effect of severe plastic deformation in equal-channel angular pressing [169–266].

Figures 5.4 and 5.5 show the creep curves of nanostructured copper and nickel in the condition immediately after the effect of severe plastic deformation (curve 1) and subsequent pre-recrystallisation annealing (curve 2). The graphs indicate that the creep curves of nanostructured copper and nickel in the condition after severe plastic deformation have the form typical of logarithmic creep even at $\sigma = (0.6–0.65)\ \sigma_{0.2}$. With increase of σ above 0.65 $\sigma_{0.2}$, the creep curves showed either two creep stages (the short steady stage and the longer, accelerated stage) or a single accelerated creep stage. A different deformation behaviour is found in the creep of nanostructured copper and nickel after preliminary pre-recrystallisation annealing. Figures 5.4 and 5.5 show that after pre-recrystallisation annealing at a temperature of 353 and 398 K, the creep curves of nanostructured copper and nickel show three stages: the non-steady, steady-state and accelerated stage. For the same value of σ, the steady-state creep rate after this type of annealing is considerably higher than immediately severe plastic

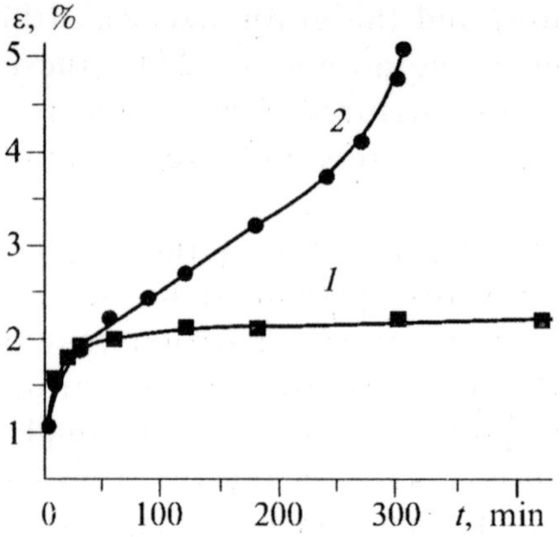

Fig. 5.4. The creep curves of specimens of nanostructured copper in the initial condition after severe plastic deformation (*1*) and after preliminary annealing at T = 398 K (2), T_{test} = 423 K, = 130 MPa.

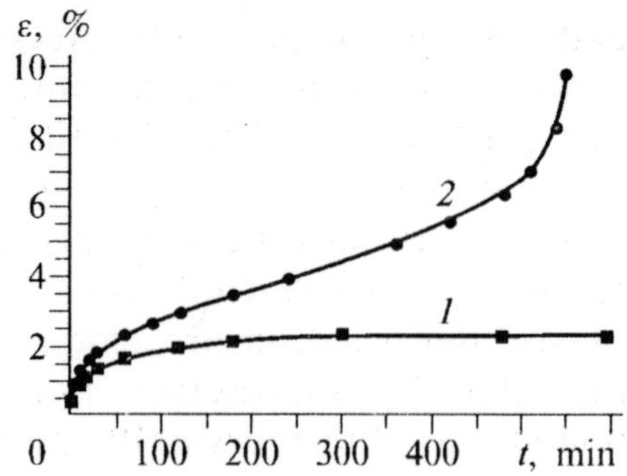

Fig. 5.5. Creep curves of specimens of nanostructured nickel in the initial condition after severe plastic deformation (1) and after preliminary annealing at T = 398 K (2); T_{test} = 423 K, = 630 MPa.

deformation. The change in the form of the creep curves and the steady stage creep rate of nanostructured copper and nickel as a result of pre-recrystallisation annealing, which is not accompanied by changes in the grain size, indicates the important role played by the condition of the grain boundaries in the formation of structure-sensitive properties of nanostructured materials. In this case, this is demonstrated for creep.

More detailed investigations of the creep of nanostructured nickel (in the condition after annealing at T = 398 K for 1 h), carried out in the temperature range 398–473 K, show that the dependence of the strain rate in the steady stage on stress for all examined temperatures is described by the well-known equation [267]:

$$\dot{\varepsilon} = A\sigma^n \exp(-Q_n / RT), \tag{5.4}$$

where $\dot{\varepsilon}$ is the steady-state creep rate, n is the exponent at the stress index, A is a constant, Q_n is the apparent activation energy of creep, and the meaning of R and T is well-known.

The dependences $\dot{\varepsilon}(\sigma)$ (Fig. 5.6) in the double logarithmic coordinates at test temperatures of 398, 423 and 443 K have the same angle of inclination with the exponent n ~ 1, which is slightly higher than the value n ~ 7 [207] for coarse-grained nickel in the same temperature and creep rate range. At the same time, at a temperature of 473 K, the lg $\dot{\varepsilon}$ –lg σ curves for nanostructured

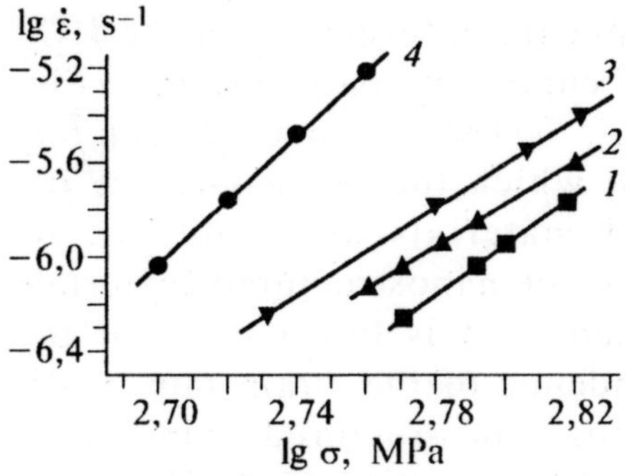

Fig. 5.6. Dependence of the steady-state creep rate on applied stress for nanostructured nickel. *1*) T = 398 K, *2*) 423, *3*) 443, *4*) 473 K.

nickel are characterised by the value n ~ 14, which is twice the value for coarse-grained nickel. The increase of n is evidently associated with the change in the ratio of the contributions of the acting mechanisms of deformation of nanostructured nickel when the temperature is increased to 473 K as a result of the recovery of the structure of the grain boundaries and, consequently, changes of their properties.

An interesting example of the deformation behaviour in the creep process is nanostructured titanium produced by severe plastic deformation using the method of equal-channel angular pressing (d ~0.3 μm). After equal-channel angular pressing, the creep curves of this material show three stages: non-steady, steady and accelerated stages (Fig. 5.7). This difference between the creep curves of nanostructured titanium and the corresponding curve for nanostructured copper and nickel is evidently associated with the

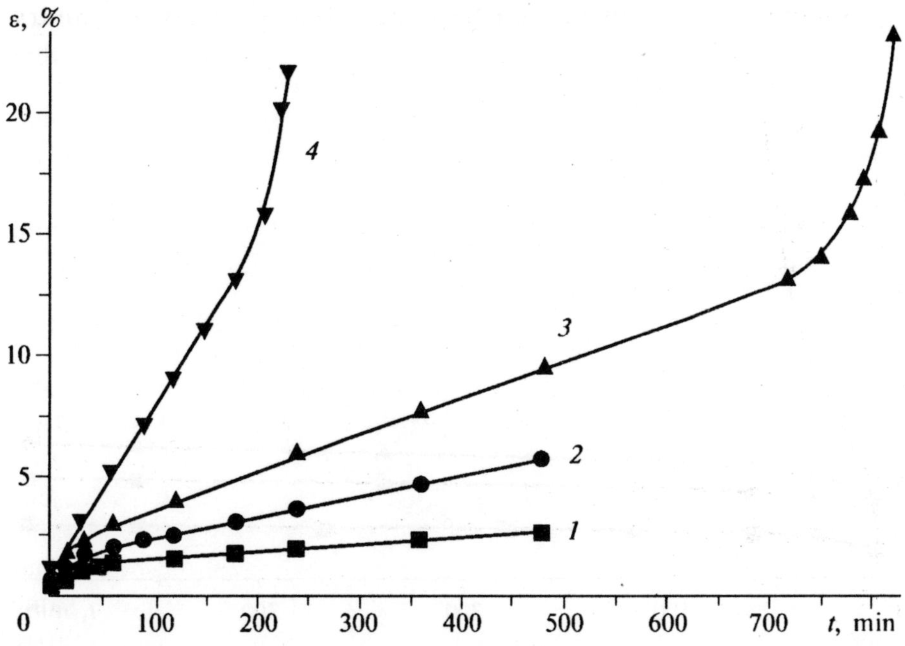

Fig. 5.7. Creep curves of specimens of nanostructured titanium VT1–0. *1*) 220, *2*) 230, *3*) 260, *4*) 300 MPa; T_{test} = 623 K.

special features of development of plastic deformation in the HCP lattice. The comparison of the creep curves of coarse-grained (Fig. 5.8) and nanostructured titanium (Fig. 5.7) shows that for each test temperature there is a stress (σ_k) at which the creep rates of the coarse-grained and nanostructured materials are identical. At stresses lower than σ_k, the creep rate of nanostructured materials is higher, and at stresses higher than σ_k it is lower than for the coarse-grained materials. The dependence of the strain rate in the steady stage of creep on the stress for nanostructured titanium, as for nanostructured nickel, is described by equation (5.4). However, the values of $n \sim 7.5$ for nanostructured titanium for the investigated temperatures of 573 and 623 K are slightly lower than the appropriate values for the coarse-grained material, $n \sim 9.5$ (Fig. 5.9).

These results indicate that the value of n for nanostructured materials is several times higher than the value predicted by theory [261,263] but differs only slightly from the experimentally determined values of n for coarse-grained materials in the given temperature range.

It is well-known that in transition of the alloys to the superplastic conditions there is a change in the mechanism of the formation from intragranular dislocation slip (IDS), controlled by volume diffusion, to grain boundary sliding (GBS) controlled by grain boundary

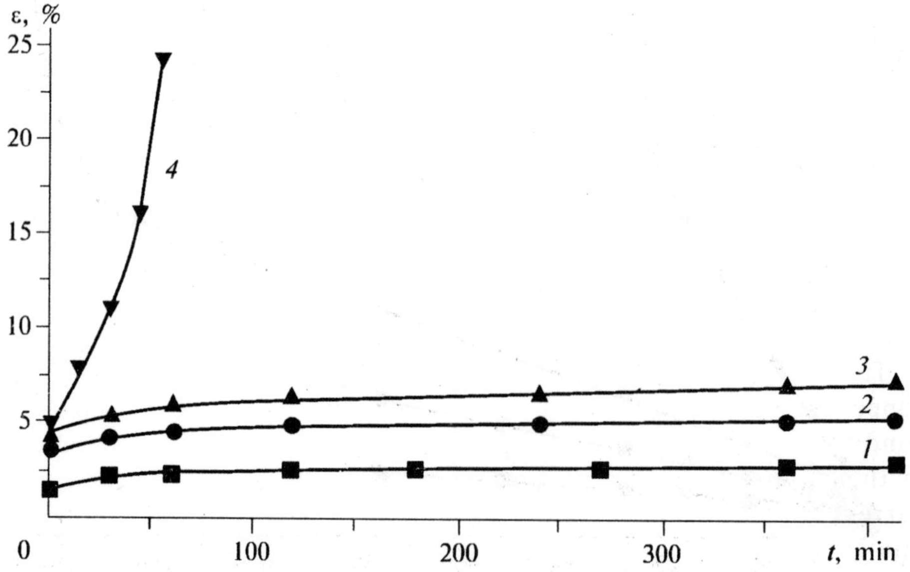

Fig. 5.8. Creep curves of specimens of coarse-grained titanium VT1–0 *1*) = 210, *2*) 220, *3*) 230, *4*) 240 MPa; T_{test} = 623 K.

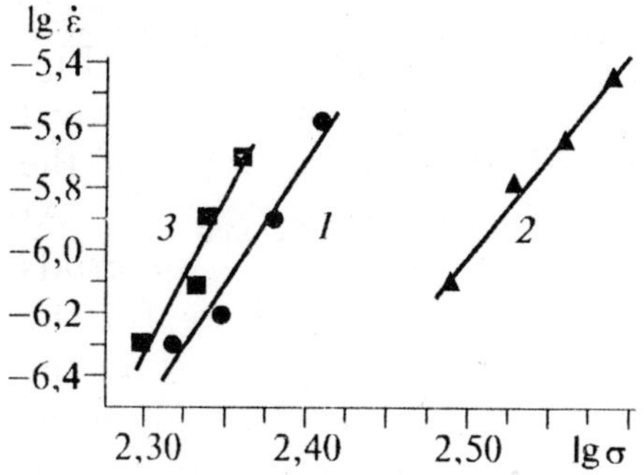

Fig. 5.9. Dependence of the steady-state creep rate on applied stress for nanostructured (*1* and *2*) and coarse-grained (*3*) titanium at T = 573 K (*2*) and 623 K (*1* and *3*).

diffusion. In this case, the apparent activation energy of creep decreases from the values equal to the activation energy of volume self-diffusion (Q_v) to the values close to the activation energy of grain boundary self-diffusion (Q_b). On the other hand, it was shown in [146, 260 of, 269] that the contribution of grain boundary sliding to the total deformation can be increased without changing the grain size, for example, by transferring the grain boundaries to the non-equilibrium (high-energy) state. This is also accompanied by a decrease of the value of Q_c to the values close to those corresponding to the activation energy of grain boundary diffusion.

Therefore, it is interesting to determine the activation energy of creep of metals and alloys in the nanostructured condition, formed by the effect of severe plastic deformation, in which are the formation of ultrafine grains is accompanied by the nonequilibrium condition of the grain boundaries.

The measurements of the apparent activation energy of creep (Q_c) were taken by the authors using the standard method of stepped variation of temperature by 10°. This included the determination of the variation of the creep rate in the steady stage. The following equation was used to the calculations [267]:

$$Q_c = R\ln(\dot{\varepsilon}_2/\dot{\varepsilon}_1)/(1/T_2/1/T_1), \qquad (5.5)$$

where T_1 and T_2 are absolute temperatures; $\dot{\varepsilon}_1$ and $\dot{\varepsilon}_2$ are the steady-state creep rates prior to and after temperature changes; R is the gas constant.

Table 5.1 shows the apparent activation energy of creep of copper, nickel and titanium in the nanostructured and coarse-grained conditions. It may be seen that the value of Q_c for the materials in the nanostructured condition is 1.5–2.5 times lower than the corresponding values for the coarse-grained materials. Taking this into account, it may be assumed that the decrease of Q_c of the investigated nanostructured materials in the temperature range (0.2–0.3) T_m in comparison with the appropriate values for the coarse-grained materials is associated with the activation of grain

Table 5.1. The apparent activation energy of creep of copper, nickel and titanium in nanostructured and coarse-grained conditions (Q_c ± 15 kJ/mole).

Material	T_t	Condition	
		Coarse-grained	Nanostructured
Copper	423	100	70
Nickel	423–473	276	115
Titanium	623	366	132

boundary sliding, controlled by grain boundary diffusion, and by the contribution of this process to total deformation.

The possibility of grain boundary sliding providing a significant contribution to the total deformation of nanostructured metals at relatively low temperatures was previously observed in [199] on the example of copper where the contribution of grain boundary sliding to the total strain in the tensile loading was approximately 20% at room temperature. The confirmation of the intensive development of grain boundary sliding in the creep of the nanostructured nickel, examined by the authors, including after preliminary annealing at T = 390 it K, is the formation of distinctive steps on the pre-polished surface of the specimen associated with sliding on the grain boundaries (Fig. 5.10), after creep at T = 423 K. It is natural to assume that at higher temperatures grain boundary sliding is also realised during creep. The significant contribution of grain boundary sliding to the total strain in active tensile loading and creep of the materials is evidently associated with two factors: firstly, with the non-equilibrium (high-energy) condition of the grain boundaries, formed by the effect of severe plastic deformation and, secondly, with the small (submicron) grain size [199, 270].

The attempt to divide the effect of these factors on the value of Q_c was made by the authors on the example of examination of the relationships governing the creep of nanostructured nickel produced by severe plastic deformation using the method of equal-channel angular pressing. The structure of the nickel, used in the experiments, after equal-channel angular pressing was described in detail in section 4.1 [191].

Figure 5.11 shows the dependences of the mean grain size and the apparent activation energy of creep on the temperature of preliminary annealing. The results show that the Q_c value of

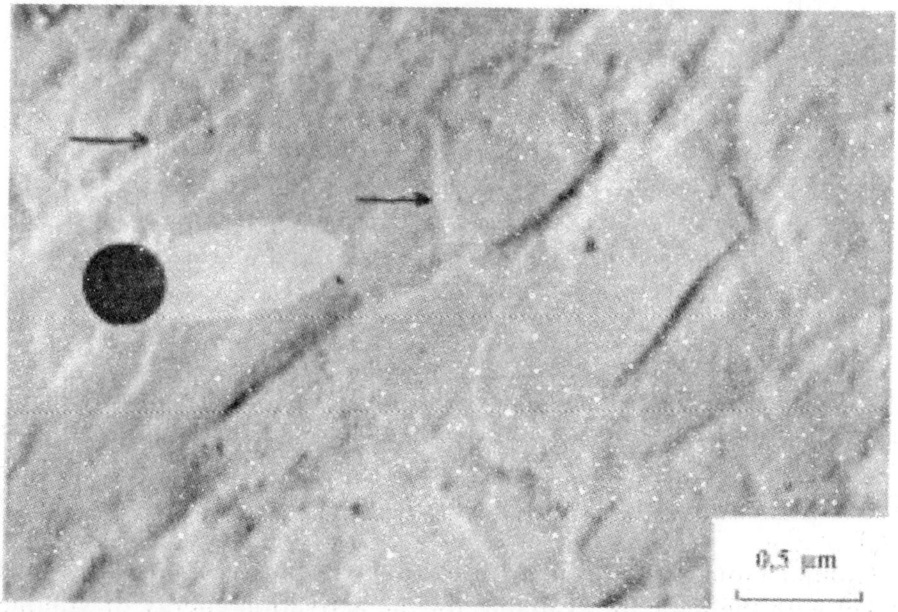

Fig. 5.10. Traces of grain boundary sliding of carbon replicas from the surface of the specimen of nanostructured nickel in creep at T = 423 K and σ = 660 MPa; the arrows indicate steps on the grain boundaries.

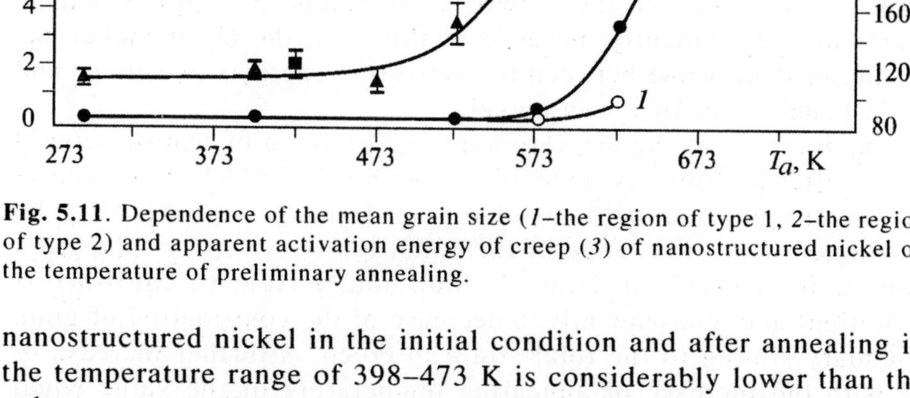

Fig. 5.11. Dependence of the mean grain size (*1*–the region of type 1, *2*–the region of type 2) and apparent activation energy of creep (*3*) of nanostructured nickel on the temperature of preliminary annealing.

nanostructured nickel in the initial condition and after annealing in the temperature range of 398–473 K is considerably lower than the value corresponding to coarse-crystalline nickel (276 kJ/mole [267]) and is equal to 115 ± 15 kJ/mole. After annealing at temperatures higher than 473 K the value of Q_c starts to increase and at the annealing temperature of 673 K, corresponding to complete recrystallisation of the nanostructure, reaches the value 270±10 kJ/mole. This value of Q_c is close to the previously given value of the apparent activation energy of creep of coarse-crystalline nickel.

Thus, the two special features in the variation of the apparent activation energy of creep of nanostructured nickel in relation to the temperature of preliminary annealing are: firstly, the decrease of Q_c for the nanostructured nickel by a factor of 2.5 in comparison with coarse-crystalline nickel; secondly, the large increase of Q_c after annealing in the temperature range 473–573 K which is not accompanied by any change of the mean grain size.

It is well-known that the main mechanism of deformation in the creep of coarse grain nickel in the investigated conditions is intragranular grain boundary slip, controlled by volume self-diffusion [207, 267]. In this case, the apparent activation energy of creep is close to the activation energy of volume self-diffusion of nickel (Q_v).

Comparison of the resultant values of Q_c for nanostructured nickel with the data are available for the activation energy of volume (Q_v) of diffusion of coarse-grained nickel (284 kJ/mole) and the activation energy of grain boundary diffusion of nanostructured nickel (Q_b*= 60.2 kJ/mole [168]) shows that the apparent activation energy of creep of nanostructured nickel is considerably lower than Q_v, but twice higher than the value of Q_b^*. Consequently, it may be assumed that in the creep of nanostructured nickel in the investigated conditions, there are two mechanisms of deformation: intracrystalline grain boundary slip, controlled by volume diffusion, and grain boundary sliding controlled by grain boundary diffusion. The contributions of these two mechanisms are approximately identical and, evidently, because of this fact, the Q_c of nickel has an intermediate value between the activation energies of volume and grain-boundary diffusion in nickel.

The increase of the activation energy of creep of nanostructured nickel after preliminary annealing at $473 < T < 573$ K, observed in the investigations, with no changes of the grain size, may be associated with the change of the condition (recovery) of the grain boundaries (transition from the non-equilibrium to equilibrium condition) and, consequently, a decrease of the contribution of grain boundary sliding to the total strain in creep. A further increase of Q_c with the increase of annealing temperature of the value equal to the activation energy of volume self-diffusion is associated with the recrystallisation and growth of the grains which, as is well-known, reduces the contribution of grain boundary sliding to the total strain and the controlling role of intragranular deformation which depends in the examined temperature range of the movement of this locations, controlled by volume self-diffusion [267, 271].

Thus, the previously examined experimental data on the relationships of creep of the nanostructured metals, produced severe plastic deformation, indicate a number of special features of their deformation behaviour in relation to the coarse-grained condition. In particular, it is the high creep resistance in the condition immediately after severe plastic deformation. Secondly, the large decrease of the creep resistance after pre-recrystallisation annealing, leading to the tempering of the nonequilibrium structure of the grain boundaries. Thirdly, the low values of the apparent activation energy of creep determined, evidently, by an increase of the contribution of grain boundary sliding to the total strain in the creep of metals in the nanostructured condition.

In addition, the fact that the values of the creep rate in the steady stage and of the apparent activation energy of creep may greatly change in relation to the condition of the grain boundaries for the same grain size, indicates that it is necessary to take into account this fact when examining the relationships governing the creep of nanostructured materials. Evidently, the contradictions in the literature data on this problem may be associated with the absence of control of the structural condition of the grain boundaries.

Composites

In the previous sections, it was noted that the presence, in the nanostructured metals, of the particles of hardening phases of oxides or carbides with the size in the nanorange (10–20 nm) greatly increases the thermal stability of their structure and mechanical properties. The low solubility of the hardening phases and their high resistance to coalescence processes make it possible to assume that the nanostructured composites have a stable structure and a high level of the mechanical properties in the conditions of the simultaneous effect of temperature and load, for example, in the conditions of creep at elevated temperatures.

Using a Cu–0.5 wt% Al_2O_3 composite as an example, the authors of [228, 229, 247, 272, 273] investigated the deformation behaviour in the creep of ultrafine-grained composites based on copper.

The initial material for the investigations was the extruded

Cu–0.5 wt% Al_2O_3 composite (Clid Cop Al–25), produced by internal oxidation (condition 1). The condition of the copper matrix was changed by additional deformation by rolling to 70% at room temperature (condition 2) and by the effect of severe plastic torsional deformation (the true logarithmic strain ~5.5) (condition 3 – nanostructured composite). The results were compared with the deformation behaviour of pure (99.98 %) copper in the nanostructured condition, produced by the effect of severe plastic deformation in torsion (d ~ 0.3 μm) and in the coarse-grained condition.

The Cu–0.5 wt% Al_2O_3 composite in the conditions 1 and 2 was characterised by the presence of a fine-grained structure with the grain size in the range 0.5–1 μm. The grain size of the copper matrix of the investigated nanostructured composites was slightly smaller, 0.2–0.3 μm. In all three conditions, the dispersed particles of the Al_2O_3 phase with the size of 15–20 nm were uniformly distributed in the volume of the material. There were regions containing larger particles (50–100 nm). As reported in the previous section, the structure was thermally stable up to T = 673 K when the first recrystallised areas appeared in the matrix.

Figure 5.12 shows the creep curves of pure copper in the condition after severe plastic torsional deformation and of the Cu–0.5 wt% Al_2O_3 composite in the conditions 1–3 at a temperature of 423 K and a stress of 225 MPa (0.6 $\sigma_{0.2}$ for the Cu–0.5 wt% Al_2O_3 composite). It may be seen that of the creep curve of pure copper in these conditions of testing shows only the accelerated

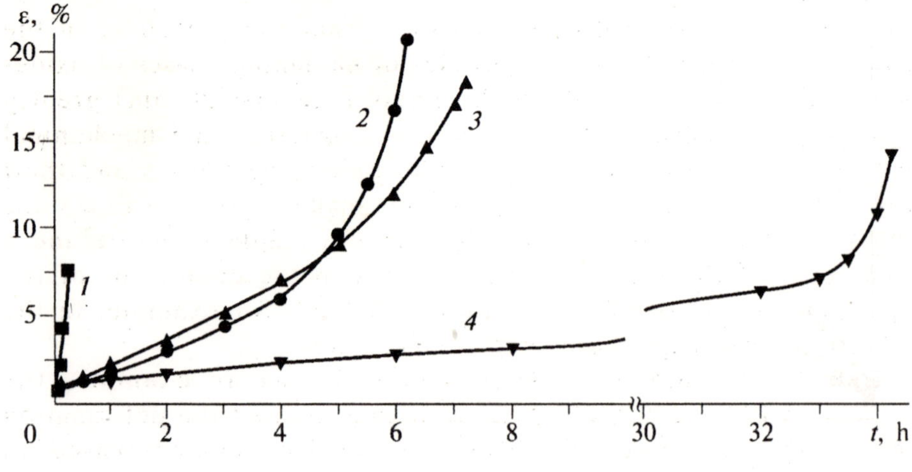

Fig. 5.12. Creep curves of the specimens of nanostructured copper (*1*) and the Cu–0.5 wt% Al_2O_3 composite in the conditions 1–3 (curves 2–4, respectively) at T = 423 K, = 285 MPa.

creep stage. The creep curves of the Cu–0.5 wt% Al_2O_3 composite are characterised by three stages of creep: non-steady, steady-state and accelerated. The steady-state creep curves of Cu–0.5 wt% Al_2O_3 composites in the conditions 1 and 2 differ only slightly ($\dot{\varepsilon}_1 = 3.8 \cdot 10^{-6}$ s^{-1}, $\dot{\varepsilon}_2 = 4.4 \cdot 10^{-6}$ s^{-1}), whereas the creep rate in the steady stage of the investigated nanostructured composite, produced by severe plastic deformation in torsion, is $\dot{\varepsilon} = 4.3 \cdot 10^{-7}$ s^{-1}, i.e. an order of magnitude smaller in comparison with the conditions 1 and 2. The strain to fracture decreases by 25%. It may be assumed that the large increase of the creep resistance of the investigated material after the effect of severe plastic deformation is associated with the special features of the produced nanostructure, characterised by the ultrafine grain size in the non-equilibrium high-energy condition of the grain boundaries (the presence of elastic stresses and high density of crystal structure defects at the grain boundaries and in the boundary regions).

This assumption has been confirmed by the results of creep tests of the Cu–0.5wt% Al_2O_3 nanostructured composite after preliminary pre-recrystallisation annealing at a temperature of 353–523 K (Fig. 5.13) which, as already mentioned, does not change the grain size but results in tempering of the defective structure of the grain boundaries. Figure 5.13 shows that the steady-state creep rate of the nanostructured Cu–0.5 wt% Al_2O_3 composite after preliminary annealing increases 1.4 times. With increase of annealing temperature, the steady state creep rate of the nanostructured composite increases and after annealing for 1 h at a temperature

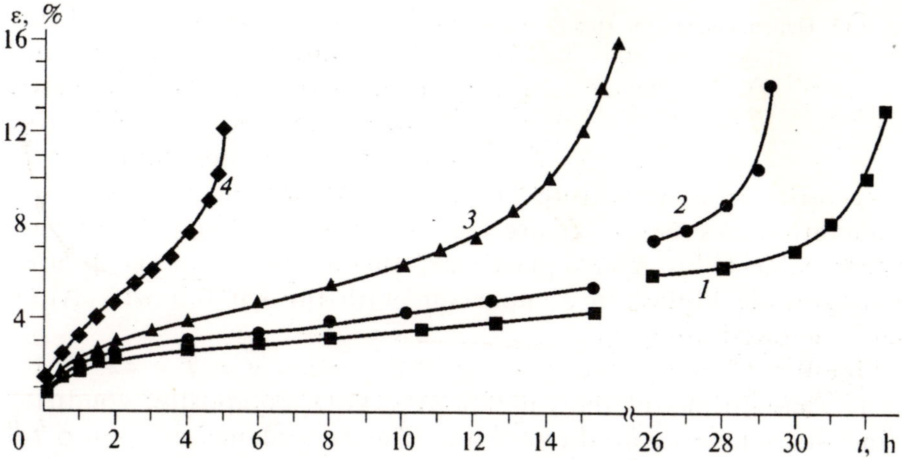

Fig. 5.13. Creep curves of the specimens of nanostructured composite Cu–0.5 wt% Al_2O_3 immediately after severe plastic deformation by torsion (1) and subsequent annealing at 353 K (*2*), 423 (*3*) and 523 K (*4*); T_{test} = 423 K, σ = 285 MPa.

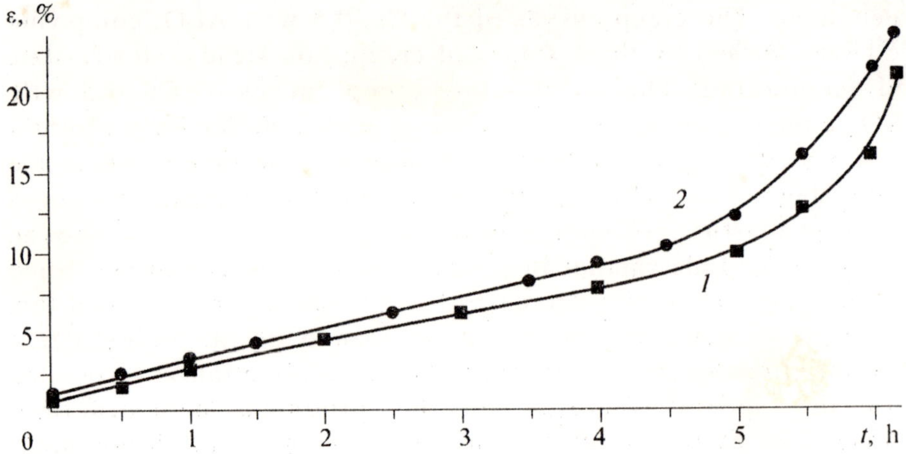

Fig. 5.14. Creep curves of the specimens of composite Cu–0.5 wt% Al_2O_3 in the condition 1 (*1*) and subsequent annealing at 523 K, 1 h (2); T_{test} = 423 K, σ = 285 MPa.

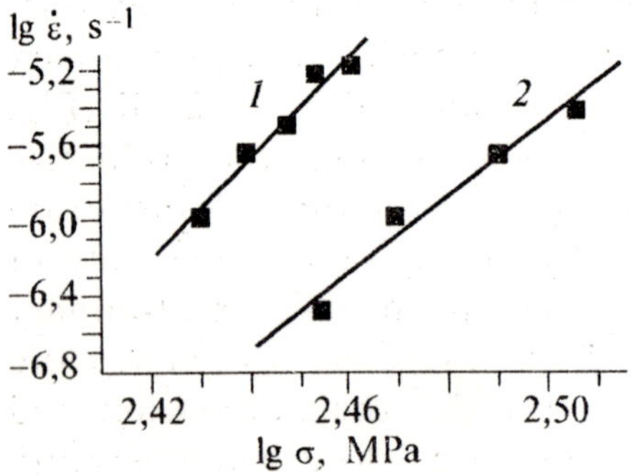

Fig. 5.15. Dependence of the steady-state creep rate of applied stress for the Cu–0.5 wt% Al_2O_3 composite in the condition 1(*1*) and 3 (*2*): T_{test} = 423 K.

of 523 K it becomes comparable with the creep rate of the Cu–0.5 wt% Al_2O_3 composite in condition 1. At the same time, preliminary annealing at a temperature of 523 K for 1 h of this composite in condition 1, characterised by a more equilibrium structure, has almost no effect of its steady state creep rate (Fig. 5.17). As already mentioned, the steady state creep rate of pure copper, subjected to severe plastic deformation in torsion, also increases after preliminary pre-recrystallisation annealing (Fig. 5.4). However, the effect of acceleration of creep after pre-recrystallisation annealing for pure copper, subjected to severe plastic deformation in torsion, is an order of magnitude higher in comparison with the Cu–0.5 wt% Al_2O_3 nanostructured composite.

Figure 5.15 shows the lg $\dot{\varepsilon}$ – lg σ dependence at T = 423 K for the two conditions of the Cu–0.5 wt% Al_2O_3 composite: condition 1 and the nanostructured condition. The dependence of $\dot{\varepsilon}$ on σ for both conditions is described by the equation of type (5.4).

It is well-known [146] that the equation of type (5.4) satisfactorily describes the steady-state creep rate of the dispersion-

Table 5.2. The apparent activation energy of creep of copper and the Cu-0.5 wt% Al_2O_3 composite

Material (condition)	Test temperature	$Q_c \pm 15$, kJ/mol
Nanostructured copper	423	71
Nanostructured copper	423	102
Composite Cu–0.5 wt.% Al_2O_3 (1)	423	71
	773	203
Composite Cu–0.5 wt.% Al_2O_3 (3)	423	127
	673	164
	773	194
Composite Cu–2.9 vol% Al_2O_3	723	286 [219]

hardened alloys. The values of n and Q_p are usually higher in comparison with pure metals (n = 3–5, $Q_p = Q_v$ is the activation energy of volume self-diffusion of metal). For example, for the Cu–Al_2O_3 dispersion-hardened alloys, containing 2.9–4.5 vol% of the hardening phase, the values of Q_p were 1.5–2 times higher than the activation energy of volume diffusion, and the value n was higher than 100 [219]. The values of n for the investigated nano-composite, determined from the angle of inclination of the curves in Fig. 5.15, for the condition 1 and the nanostructured condition are equal to the respectively 30 and 22, which is also considerably higher in comparison with the pure metals. However, the values of Q_p (table 5.2) for the investigated composite in both conditions are smaller than the activation energy of volumes of diffusion of copper (Q_v = 197 kJ/mole [207]). The comparison of the values of Q_p for the Cu–0.5 wt% Al_2O_3 composite in the condition 1 is close to the value of Q_p of copper in the nano-structured condition (71 kJ/mole) and smaller than the value corresponding to coarse-grained material (102 kJ/mole). The value Q_p of the investigated composite in the nanostructured condition is equal to 127 kJ/mole which is higher than the Q_p values of copper in the nanostructured and coarse-grained conditions but lower than the values corresponding to the Cu–Al_2O_3 coarse-grained composite.

The creep tests carried out in the temperature range 473–773 K show that the large difference in the steady state creep rates of the Cu–0.5 wt% Al_2O_3 the composite in the condition 1 and in the nanostructured condition are maintained up to a temperature of

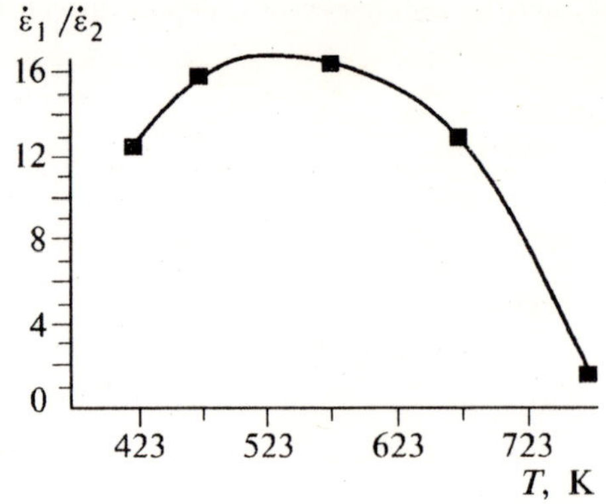

Fig. 5.16. Dependence of the value of the $\dot{\varepsilon}_1/\dot{\varepsilon}_2$ ratio/on test temperature. $\dot{\varepsilon}_1$ and $\dot{\varepsilon}_2$ are the creep rates in the steady-stage of creep of the Cu–0.5 wt% Al_2O_3 composite in the extruded and nanostructured condition respectively at the same temperature and stress.

673K. At a test temperature of 773 K, the steady state creep rates of the time to fracture in these conditions becomes similar.

Figure 5.16 shows the dependence/on the test temperature (are the steady state creep rates of the Cu–0.5 wt% Al_2O_3 composite in the conditions 1 and 3, respectively for the same temperature and stress).

Figure 5.16 shows that the rapid decrease of the creep resistance of the Cu–0.5 wt% Al_2O_3 composite in the nanostructured condition takes place in the temperature range 673–773 K. It was mentioned previously that at temperatures higher than 673 K, the Cu–0.5 wt% Al_2O_3 composite undergoes recrystallisation. It may be assumed that the decrease of the creep resistance of this composite in the nanostructured condition at temperatures above 673 K is associated with the increase of the grain size and, possibly, with a change of the deformation mechanism. The latter is confirmed by the increase of the apparent activation energy of creep of the Cu–0.5 wt% Al_2O_3 composite in both conditions up to 194–203 kJ/mole (Table 5.2), which is close to the activation energy of volume self-diffusion of copper.

Thus, the results show that the mechanisms, controlling the deformation behaviour of the nanostructured composites based on copper, produced by the effect of severe plastic deformation, in the temperature range (0.2–0.3) Tm may differ from the appropriate values not only in the case of coarse-grained dispersion-hardened copper-based alloys but also the identical ultrafine grained composites, produced by the consolidation of a mixture of powders

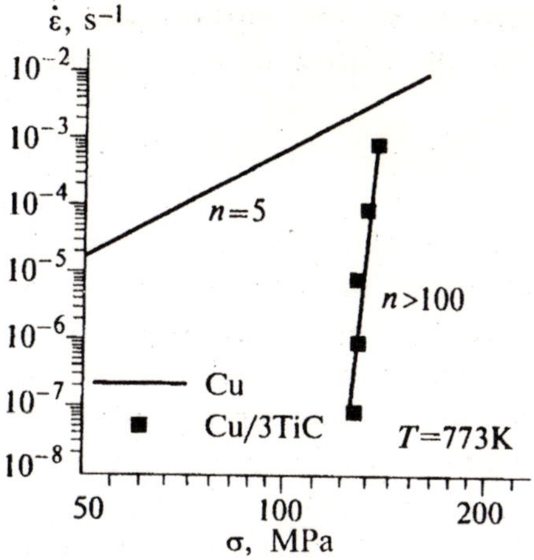

Fig. 5.17. Dependence of the creep rate of copper and the Cu–3 vol% TiC composite on stress [221].

and subsequent hot extrusion.

The authors of [221] investigated the creep mechanism of the ultrafine grained composites using copper-based composites (grain size ~500 nm), hardened by dispersed particles of TiC (d = 10–50 nm), produced by consolidation of a mixture of fine dispersion powders with subsequent hot extrusion. Figure 5.17 shows the results of tests of the dependence of the steady-state creep rate ($\dot{\varepsilon}$) of the stress (σ) for the Cu–3 vol% TiC composite at 773 K [221]. It may be seen that this dependence is described by the conventional exponential creep law of the type (5.4) with n greater than 100, which is considerably higher than the values of n = 5 for pure coarse-grained copper and is comparable with the values of n of dispersion-hardened coarse-grained materials [219].

To explain the anomalously strong dependence of the creep rate on stress of the investigated composite, the authors of [221] used the model developed by Rosler and Artz [274, 275] according to which the creep behaviour of dispersion-hardened materials is determined by the local climb and subsequent thermally activated separation of dislocations from the particles of the hardening phase. In accordance with the model [274, 275], the steady-state creep rate is described by the following equation:

$$\dot{\varepsilon} = \frac{6D_v l \rho_D}{b} \exp\left[-\frac{Gb^2 r}{kT}\left[(1-k_b)(1-\sigma/\sigma_d)\right]^{3/2}\right], \tag{5.6}$$

Here D_v is the coefficient of volume diffusion; ρ_D is the dislocation density; b is the Burgers vector; G is the shear modulus; r is the radius of the particles of the hardening phase; $k_б$ is a parameter describing stress relaxation in the vicinity of the particle–matrix surface; σ is the applied normal stress; σ_d is the separation stress of the dislocation from the particle; $2l$ is the distance between the disperse particles.

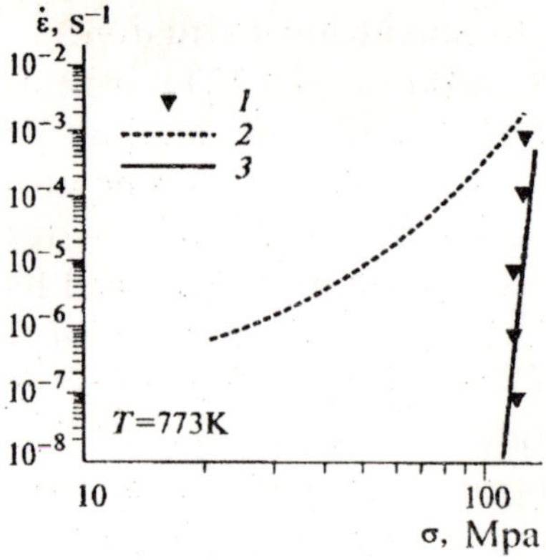

Fig. 5.18. Comparison of the models of diffusional and dislocational creep with the experimental values for the Cu–3 vol% TiC composite [221]. 1) experimental data; 2) the model of diffusional creep, k_b = 0.75; 3) the model of dislocational creep, k_b = 0.75.

To calculate σ_d, the authors of [221] use the following equation:

$$\sigma_d = \sigma_0\sqrt{1-k_b^2}, \tag{5.7}$$

where σ_0 is the Orowan stress. It was assumed that the particles of the hardening phase, distributed at the grain powders, cannot bend by the Orowan mechanism and play the role of obstacles for grain boundary sliding. This enabled the authors of [221] to calculate the Orowan stress taking into account only the fraction of the particles of the hardening phase, distributed in the body of the grains. Calculations were conducted for the Cu–3 vol% TiC composite. Figure 5.18 shows that at T = 773 K and k_b = 0.75, the experimental dependence of $\dot{\varepsilon}$ on σ for the Cu–3 vol% TiC composite is in good agreement with the dependence calculated using equations (5.6) and (5.7). According to these data, the authors of [221] concluded that at the investigated temperature compression of the ultrafine-grained composites based on copper is accompanied by dislocational creep, controlled by thermally activated separation of dislocations from the particles.

A different conclusion was obtained by the authors of [223] who investigated the dependence of the creep of ultrafine-grained composites, hardened by the dispersed nanosized particles using the Ag--5vol% Cr_2O_3 composites as an example (the grain size of silver was varied in the range of 50 nm to 8 μm). According to the above authors, the previously examined process of dislocation climb and subsequent thermal separation from the particles of the hardening phase can determine the steady-state creep rate because the threshold stress of this process $\sigma_p \cong 2$ MPa is considerably lower than the experimental values of stress of plastic flow for this composite. On the basis of analysis of experimental data, Grundmann et al proposed to modify the Rosler–Artz model by introducing into the model of the long-range a thermal component of stress (σ_p), acting on the dislocation during creep, and write the dependence of $\dot{\varepsilon}$ on σ in the following form:

$$\dot{\varepsilon} = \dot{\varepsilon}_0 \exp\left(-2T_r \left[(1-k_b)\left(1 - \frac{\sigma - \sigma_p}{\sigma_d}\right)\right]^{3/2} / kT \right), \qquad (5.8).$$

where ε_0 = 3 D_v l/M_b; M = 3 is the Taylor factor; T_r = $5.6 \cdot 10^{-10}$ N.

The remaining notations are the same as in equation (5.6). Following the classical hardening concept by Taylor, the authors of [223] proposed to calculate σ_p using the equation:

$$\sigma_p = \alpha G b M \rho_D{}^{1/2}, \qquad (5.9)$$

where α = 0.2 is the coefficient of elastic interaction.

Figure 5.19 shows the experimentally determined dependence of $\dot{\varepsilon}$–σ for pure silver and the Ag–5 vol% Cr_2O_3 composite (grain size 0.3–8 μm). For comparison, the figure shows the $\dot{\varepsilon}$–σ dependence calculated from the experimental results for the case of uniform distribution of the particles of the hardening phase Cr_2O_3 (ideal composite) in the volume, and the $\dot{\varepsilon}$–σ dependence calculated for the Ag–5 vol % Cr_2O_3 composite using equations (5.8) and (5.9) for different values of k_b (dot-and-dash lines). It may be seen that the values of the $\dot{\varepsilon}$–σ dependence, calculated using the model

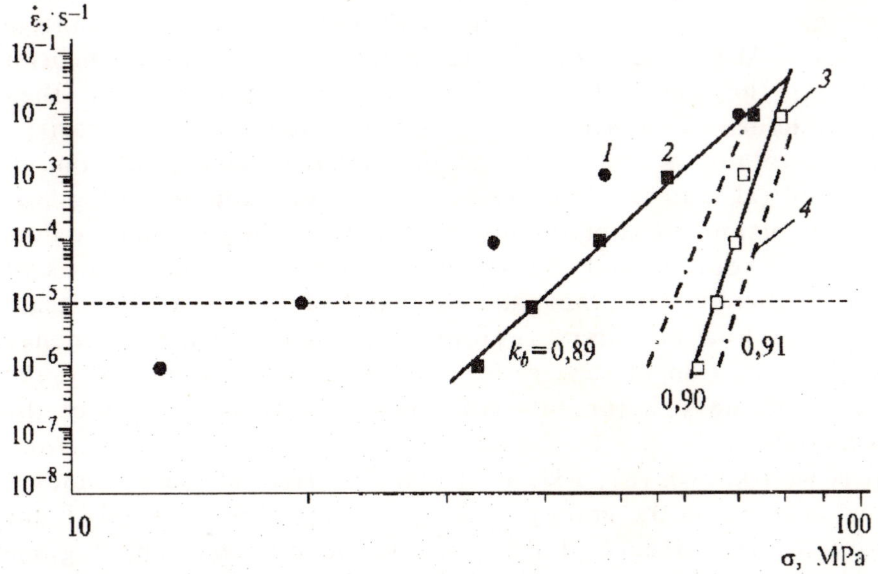

Fig. 5.19. Experimentally determined dependence of the creep rate on stress and the dependence calculated from equation (5.8) [223]. *1*) pure silver; 2) Ag–5% Cr_2O_3 composite; *3*) ideal composites; *4*) calculated relationships.

proposed in [223] for the Ag–5 vol% Cr_2O_3 composite correspond to the dependence calculated from the experimental data at k_b = 0.9.

Thus, the experimental results show that in the literature there are no united views regarding the nature of the experimentally observed special features of the creep of the ultrafine grain materials. It should also be mentioned that the examination of the relationships of creep of these materials, including nanostructured materials, produced by the effect of severe plastic deformation, were started only recently and the currently available experimental data are clearly insufficient for making any justified conclusions regarding the mechanisms determining the steady-state creep rate. It is necessary to carry out further investigations, directed to the development of the models of creep of ultrafine grain materials.

5.2. EFFECT OF ACTIVATION OF CREEP OF NANOSTRUCTURED METALS BY THE EFFECT OF GRAIN BOUNDARY DIFFUSION FLOWS OF ATOMS FROM AN EXTERNAL SOURCE

The internal interfaces (grain and interphase boundaries) are the active elements of the structure of polycrystals and have a significant effect on their important physical and mechanical properties. At elevated and high temperatures, the grain boundaries represent the path of intensive mass transfer because they are characterised by higher diffusion permittivity in comparison with the volume of the grains. In the actual service conditions of the structural materials this results in the development of grains by the diffusion flows of impurity atoms from the external medium or internal sources and the associated formation of features of the equilibrium (internal stresses and defects) at the grain boundaries and boundary regions. This may greatly influence both the condition of the grain boundaries and, consequently, the stability of the structure of structure-sensitive properties of the polycrystals.

It is well-known [82, 131, 146] that the effect of grain boundary diffusion flows on the grain boundaries in poly- and bicrystals is the reason for the effects of diffusion-induced migration of grain boundaries (DIMGB) and activation of grain boundary sliding. The activation of the latter in the specific conditions may in turn result in a large decrease of creep strength [131, 146], the formation of the superplastic condition [276] and is often accompanied by the formation of grain boundary cracks [146]. For example, the creep rate of molybdenum (grain size 15–30 μm) in the conditions of activation of grain boundaries by the diffusion flows of nickel increases by up to 102 times and deformation to fracture (plasticity) increases 10 times or more [276]. The density of the surface cracks increases by approximately 20% [277].

In [144, 146, 276–279] it was shown that the effects of DIMBG and diffusion-induced grain boundary sliding (DIGBS) occur only in the diffusion regime B1 (according to the classification proposed in [74]). This classification of the physical meaning of the regime B_1 is described in greater detail in chapter 2.

In [131, 144, 276] it was established that the grain boundary diffusion of impurities of substitutional elements in the regime B_1 is accompanied by the nucleation of individual dislocations at the grain boundaries and in the boundary areas, and by the formation of dislocation walls. It is assumed that these effects are the consequence of partial relaxation of the high-energy nonequilibrium condition of the grain boundaries with the unstable atomic structure formed under the effect of the directional diffusion flows of impurities through the grain boundaries in the regime B1. As already mentioned (chapter 2), the resistance to grain boundary sliding greatly decreases in these conditions.

It is well-known that the development of processes of grain boundary sliding and crack formation depends greatly on the type of the structure of the grain boundaries [206, 280]. Thus, the small-angle and special boundaries, characterised by low energy, are more resistant to intergranular failure in comparison with the general high-angle grain boundaries. The rate of grain boundary sliding at special boundaries at the same stress is considerably lower in comparison with the general type grain boundaries. Naturally, the deformation behaviour of the polycrystalline materials in creep in the conditions of the effect of diffusion grain boundary flows of the impurities is determined greatly by the type of grain-boundary ensemble (the ratio of the number of low- and high-energy boundaries).

The authors of [268, 281] analyzed the effect of the type of grain boundary ensemble on the development of the process of grain boundary sliding and deformation behaviour of polycrystalline materials in the conditions of diffusion of impurity atoms from the surface. Investigations were carried out on Ni(Cu) and Ni(Ag) systems (in the brackets here and further on the diffusant

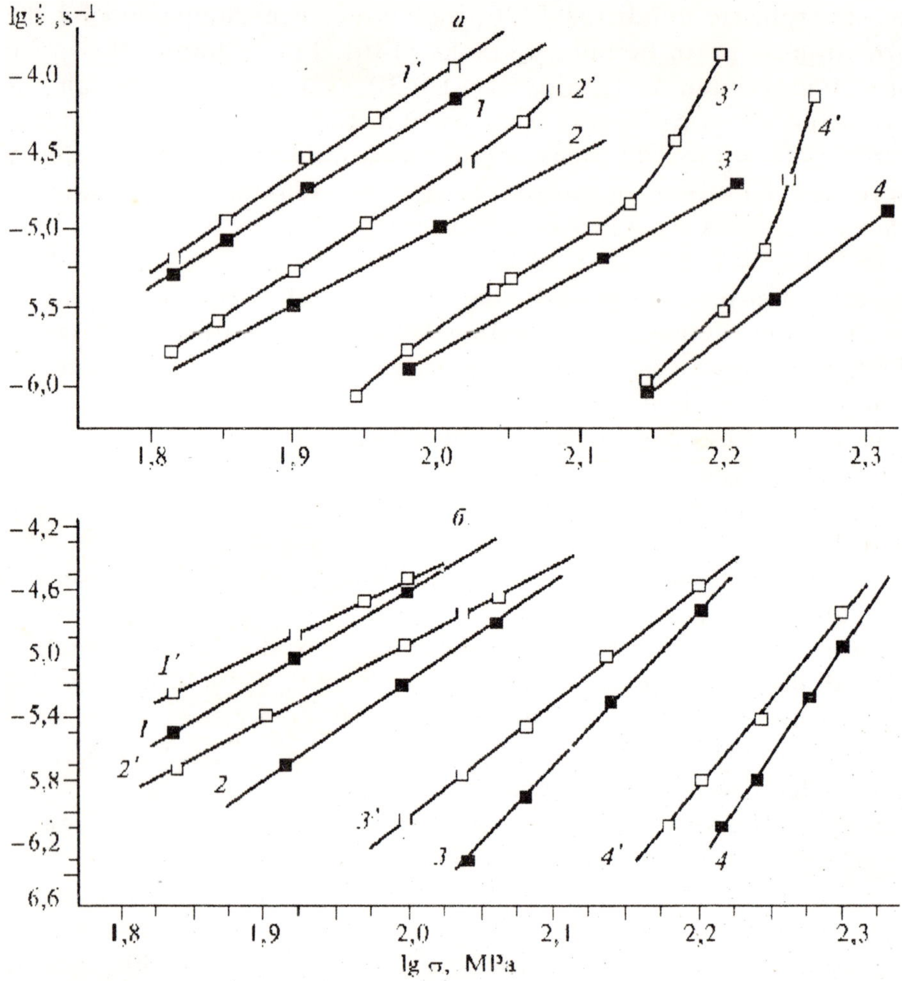

Fig. 5.20. Dependence of the steady-state creep rate on applied stress for nickel in vacuum (*1–4*) and in the conditions of diffusion of silver from the surface (1'-4') at temperatures of 973 (*1, 1'*), 923 (2, 2'), 873 (*3, 3'*) and 823 K (*4, 4'*). *a*) structure 1; *b*) structure 2.

admixture is given) in the coarse-grained (*d* ~20 μm) condition.

As a result of rolling and subsequent annealing, two types of grain boundary ensembles with different fractions of the special-type grain boundaries were produced: 5% (structure 1) and 60% (structure 2).

The calculations of the asymptotic parameters λ and *v*, characterising the diffusion regime (chapter 2), show that the more suitable temperature range for creep experiments in the diffusion regime B_1 for the system Ni(Cu) and Ni(Ag) is 773–9703 K.

Figures 5.20 *a, b*–5.22 show that in the given temperature range

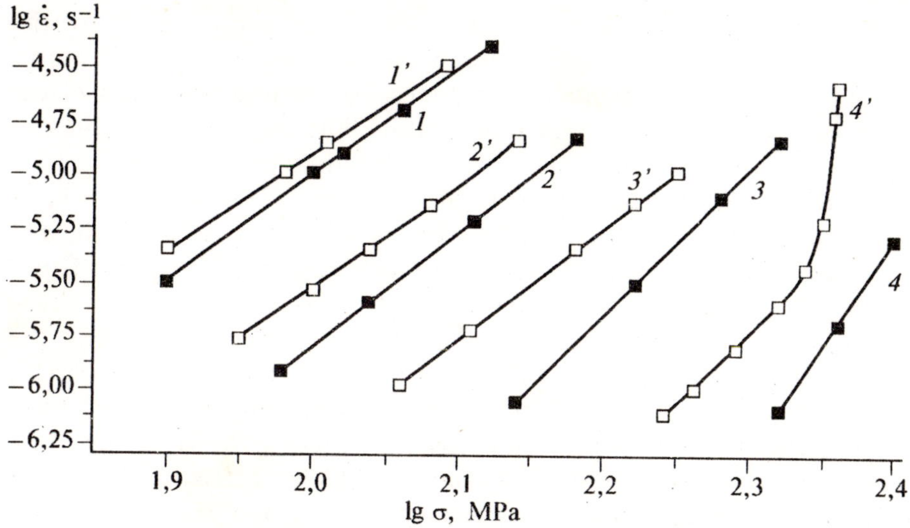

Fig. 5.21. Dependence of the steady-state creep rate of applied stress for nickel (structure 1) in vacuum (*1-4*) and in the conditions of diffusion of copper from the surface (1'–4') at temperatures of 923 (*1, 1'*), 873 (*2, 2'*), 823 (*3, 3'*) and 773 K (*4, 4'*).

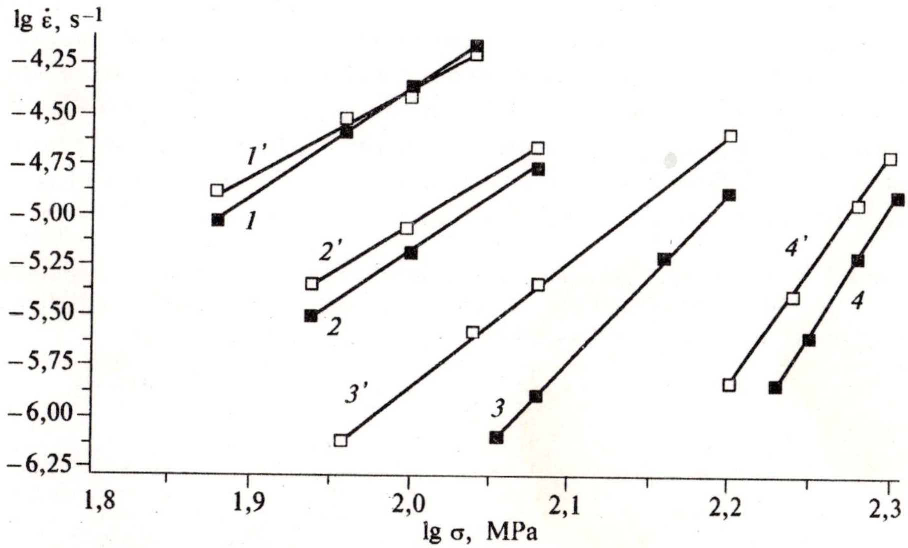

Fig. 5.22. Dependence of the steady-state creep rate on applied stress for nickel (structure 2) in vacuum (*1–4*) and in the conditions of diffusion of copper from the surface (*1–4'*) at temperatures of 973 (*1, 1'*), 923 (*2, 2'*), 873 (*3, 3'*) and 823 (*4, 4'*).

the creep rate of nickel actually increases as a result of the activation of the grain boundaries by the diffusion flows of atoms of copper and silver from the surface. The strength of this effect depends not only on the test temperature but also on the values of applied stresses and the type of grain boundary ensemble.

In nickel with the structure 1 (Fig. 5.20*a*, 5.21) the variation of the strain rate in the range 10^{-6}–10^{-4} s^{-1} causes that the acceleration effect of creep under the effect of the diffusion flows of copper and silver, determined as the ratio $\dot{\varepsilon}_2/\dot{\varepsilon}_1$ ($\dot{\varepsilon}_1$ and $\dot{\varepsilon}_2$ are the steady-state creep rates of nickel in vacuum and in the conditions of diffusion of the impurity from the surface, respectively) increases 2–12 times depending on test temperature. The exponent at the stress index $n = d\lg\dot{\varepsilon}/d\lg\sigma$ (σ is the flow stress) in creep of nickel in the conditions of diffusion of silver from the surface initially decreases to 5 when the strain rate is increased from 10^{-6} to 10^{-4} s^{-1} and subsequently increases to 20–30 (Figs. 5.20*a*, 5.21). On the other hand, in nickel with structure 2, the strength of the effect of acceleration of creep at all test temperatures is maximum (increase of $\dot{\varepsilon}_2/\dot{\varepsilon}_1$ by a factor of 2.5–3.5) at low strain rates and monotonically decreases with increasing strain rate (Figs. 5.20*b*, 5.21, 5.22). The value of n in nickel with the structure 2 is independent of the steady-state creep rate and is determined only by the test temperature (Table 5.3, Figs. 5.20*b*, 5.21 and 5.22). The increase of the fraction of low-energy grain boundaries in the grain boundary ensemble of nickel from ~5 to ~60% reduces by a factor of 3–6 the strength of the effect of acceleration of creep associated with the influence of the grain boundary diffusion flows of silver, and shifts the maximum of the effect of the range of low strain rates.

Table 5.3. Temperature dependence of the contribution of grain boundary sliding to the total strain (γ), the exponent at the stress index (Q_n) and the apparent activation energy of creep (Q_c) in nickel and the Ni(Ag) system.

T, K	γ, %				n				$Q_n \pm 20$, [kJ/mole]			
	1		2		1		2		1		2	
	Ni	Ni (Ag)	Ni	Ni (Ag)	Ni	Ni (Ag)	Ni	Ni (Ag)	Ni	Ni (Ag)	Ni	Ni (Ag)
823	12.5	48	10.8	44	6.6	10	10	7.7	277.2	–	276.4	–
873	15	50	14.3	47	5	5	7.7	4.8	278.6	182.3	274.1	168.4
923	17	54	15.4	44	5	5.3	5.9	5	–	–	–	–
973	19.2	41	17.6	38	5.3	5.5	6.3	5	–	–	–	–

Examination of the evolution of the microstructure of the surface of the specimen shows that the contributions of the acting mechanisms of deformation in creep for pure nickel and the systems Ni(Cu) and Ni(Ag) differ. Regardless of the type of grain boundary ensemble, the forming of nickel in the entire investigated range of temperatures and rates is carried out mainly by intragranular deformation (the contribution of grain boundary sliding does not exceed 20%) (Table 5.4). The grain boundary cracks and the traces of grain boundary sliding on the surface of nickel are not detected up to the third stage of creep (at a strain higher than 20%).

The measured contribution of grain boundary sliding to the total deformation of pure nickel at a creep rate of $\dot{\varepsilon} \approx 1.5 \cdot 10^{-5}$ s^{-1} increases with increasing temperature from 9% at a temperature of 773 K to 19% at 973 K (Table 5.4). As a result of grain boundary sliding, the strain rate changes in the range $1.3 \cdot 10^{-6}$ $3 \cdot 10^{-6}$ s^{-1}. These data are in good agreement with the rate of grain boundary sliding, calculated from the equation [282]:

$$\dot{\varepsilon}_{GBS} = A\frac{D_b}{kT}\left(\frac{\sigma}{E}\right)^2\left(\frac{b}{d}\right)^2, \qquad (5.10)$$

where $A = 1.5 \cdot 10^3$ is a constant [283]; $b = 2.49 \cdot 10^{-10}$ m is the Burgers vector in nickel; $d = 2 \cdot 10^{-5}$ m is the mean grain size; D_b is the coefficient of grain boundary self-diffusion of nickel; the

Table 5.4. Temperature dependence of the total strain to fracture (ε), deformation as a result of grain boundary sliding (ε_{GBS}) and the contribution of grain boundary sliding to total strain (γ) for nickel and the systems Ni(Cu) and Ni (Ag) in steady-state creep with the rate of $\dot{\varepsilon} \approx 1.5 \cdot 10^{-5}$ s^{-1}

T, K	ε			ε_{GBS}			γ		
	%								
	Ni	Ni (Cu)	Ni (Ag)	Ni	Ni (Cu)	Ni (Ag)	Ni	Ni (Cu)	Ni (Ag)
773	28	23	–	2.5	10	–	8.9	43	–
823	32	28	25	4	11	12	12.5	39	48
873	40	36	32	6	14	16	15	39	50
923	47	44	37	8	18	20	17	41	54
973	57	–	51	11	–	21	19.2	–	41

remaining notations are generally known. This equation is valid for the case in which the accommodation of the grain boundary sliding takes place as a result of deformation of the adjacent grains and is controlled by the process of grain boundary diffusion. For a temperature of 973 K and the applied stress of σ = 282 MPa corresponding to the given creep rate (E = 150 GPa; $D_b = 2.3 \cdot 10^{-13}$ m²/s), the rate of grain boundary sliding, calculated from equation (5.0), is ~$1.1 \cdot 10^{-6}$ s^{-1}.

The strain rate in the creep of pure nickel can be estimated using the equation derived in [207], for the creep controlled by the processes of dislocation sliding and climb:

$$\dot{\varepsilon} = \frac{A_s DGb}{kT}\left(\frac{\sigma}{G}\right)^n, \tag{5.11}$$

where $D = D_V + 10\left(\frac{\sigma}{G}\right)^2 \frac{a_c D_c}{b^2}$ is the effective coefficient of diffusion;

$A_s = (\sqrt{3})^{n+1} A_{Ni}$ ($A_{Ni} = 3 \cdot 10^6$ is Dorn's constant for nickel); a_c is the cross-sectional area of the dislocation kernel; D_c is the coefficient of diffusion through the dislocations; the remaining notations correspond to those in equation (5.10). For the investigated temperature range $n = 7$, $a_c D_{oc} = 3.1 \cdot 10^{-23}$ m⁴/s, the activation energy of diffusion through dislocation pipes Q = 170 kJ/mole [207]. The values of lg $\dot{\varepsilon}$ calculated from this equation for the stress of σ = 159 MPa at temperatures of 773–973 K are equal to ~–8.22 ÷ –5.6. They are in good agreement with the values of lg $\dot{\varepsilon}$ for the temperatures of 773–823 K (–8 ÷ –5.5), obtained in the experiments. The value of n is ~6.6 (Fig. 5.20). With increasing temperature (873–973 K) values of n decrease to 5, whereas the value of lg $\dot{\varepsilon}$ increases to –3. This decrease of the values of n in comparison with the values reported in [207] and also the increase of the strain rate in comparison with the value calculated from equation (5.11) is evidently associated with the increase of the contribution of grain boundary sliding to total strain (Table 5.4). (For the process of grain boundary sliding in accordance with (5.10) n = 2).

In contrast to pure nickel, the deformation in creep in the systems Ni (Cu) and Ni (Ag) is associated with the development of the processes of grain boundary sliding and crack formation. The contribution of grain boundary sliding to the total strain increases

Table 5.5. Rate dependence of the apparent activation energy of creep (Q_c), the contribution of grain boundary sliding to total strain (γ) and the density of grain boundary cracks (ρ_T) in the Ni(Ag) system at T = 873 K

ε, s^{-1}	$Q_c \pm 20$, kJ/mol		γ, %		ρ_T, cm^2	
	1	2	1	2	1 $\varepsilon \approx 20\%$	1 $\varepsilon \approx 20\%$
$2.4 \cdot 10^{-6}$	168.4	182.3	56	50	184	52
$5.6 \cdot 10^{-6}$	148.6	178.4	53	51	386	74
$7.1 \cdot 10^{-6}$	132.2	165.7	52	51	442	86
10^{-5}	112.4	171.3	50	49	670	123

in comparison with pure nickel by a factor of 2–4 (Table 5.4). The first grain boundary cracks appear in these systems already at a strain of ~6–15%, depending on the temperature and creep rate and at the end of the steady stage are found throughout the entire surface of the specimen. The formation of cracks at of the grain boundaries in these systems is accompanied by the formation of breaks and displacement of the grooves indicating the development of the process of grain boundary sliding. The contribution of grain boundary sliding to total strain increases to 40–50% (Table 5.4). However, at the same rate and strain, the density of the grain boundary cracks of the surface of the specimens with structure 2 is 4–5 times smaller in comparison with the specimens with the structure 1 detected only at general-type grain boundaries (Table 5.5).

The special features of the structure of nickel, in which the grain boundary ensemble contains up to 60% of special-type grain boundaries, is the large increase (by a factor of 2–3) of the mean grain size after creep. (Identical coarsening of the structure 2 is also found after creep in the conditions of diffusion of the solute from the surface). However, the treatment of the surface of the specimens in a chemical etching agent, developing mainly the special boundaries, shows that this coarsening of the structure is only apparent. In fact, the group of grains (usually 2–4 grains), separated by special-type boundaries, behaves during deformation as a single larger grain with the general boundaries characterised by grain boundary sliding on the formation of grain boundary cracks. Evidently, this is caused by the fact that special boundaries are a small obstacle to the movement of dislocations and sliding through the boundaries is easily transferred from grain to grain [206].

The measurements of the apparent activation energy of creep, carried out by the methods of the temperature jump for nickel and the Ni (Cu) and Ni (Ag) systems in the investigated temperature ratio of measurement of the apparent activation energy of creep show that Q_p of pure nickel is independent of the type of grain boundary ensemble and for the investigated temperature range (773–973 K) it is equal to 278 ± 20 kJ/mole. This value of Q_p is close to the activation energy of volume self-diffusion of nickel (Q_v = 224 kJ/mole [207]) indicating that deformation of nickel in these conditions is controlled by volume diffusion. In the system Ni(Cu) and Ni(Ag), the Q_p value of nickel for both types of grain boundary ensemble decreases in comparison with pure nickel. In this case, the value of Q_p for the nickel with the structure 2 is almost completely independent of the creep rate, and in the case of nickel with structure 1 decreases with increasing strain rate. Table 5.5 shows that Q_p of the Ni (Cu) and Ni(Ag) systems is situated between the values for volume Q_v and grain boundary Q_b = 115 kJ/mole [207] of diffusion of nickel.

On the basis of these results it may be assumed that in the creep of nickel in the condition of diffusion of solutes from the surface, two deformation mechanisms operate regardless of the type of grain boundary ensemble: intragranular dislocation slip (IDS), controlled by volume diffusion, and grain boundary sliding, controlled by grain boundary diffusion. The contributions of these two mechanisms are approximately equal and the apparent energy of activation of creep occupies an intermediate position between the values of the activation energy of the volume and grain boundary self-diffusion.

The comparison of the variation of Q_p at the increase of the strain rate with the variation of the contribution of grain boundary sliding to total strain (γ) shows that for almost the same and constant value of γ in nickel with the structure 1, the value of Q_p decreases in the investigated rate range by almost 30%, whereas in the nickel with the structure 2 it remains constant (Table 5.5). It is well-known [156, 284] that the diffusion along the grain boundaries with cracks may be identical to that occurring on the surface of materials. In this case, the activation energy of grain boundary diffusion may decrease to the values close to surface diffusion. Consequently, it may be assumed that the decrease of Q_p in the creep of nickel in the conditions of grain boundary diffusion of silver when the surface contains a large number of cracks, may be determined not only by the increase of the contribution of grain boundary sliding, controlled by grain boundary diffusion, to the total strain, but also by the decrease of the activation energy of grain boundary diffusion as a result of the formation of microcracks at the grain boundary. Consequently, the existence of the rate of dependence of Q_p in nickel with the structure 1 and the absence of such a trend in the nickel with structure 2 may be explained by the considerably higher density of grain boundary cracks in the structure 1 in comparison with structure 2 and by the variation of this density with the strain rate (Table 5.5). This assumption also determines the good agreement of the values of the depth of penetration of silver into nickel, determined by ???, and the depth of penetration of cracks into nickel. For example, in the process of creep at T = 873 K and σ = 135 MPa for 3 hours, silver penetrates into nickel with structure 1 to a depth of 65 μm, and the depth of propagation of the cracks is 80 μm. In the same conditions, silver penetrate into the nickel with the structure 2 to a depth of 30 μm. The depth of penetration of cracks is approximately the same.

The results of metallographic examination also showed that grain-boundary flows of copper and silver from the surface in the process of creep activated in nickel not only grain boundary sliding but also migration of the grain boundaries. Table 5.6 shows the data on the variation of the rate of migration of the grain boundaries (V) in relation to the applied stress in nickel and the Ni(Cu) and Ni(Ag) systems for creep at a temperature of 873 K, and also the range of the grain boundary migration rates calculated for these temperature from the equation proposed in [144], in which the diffusion regime of type B1 is realised. As indicated by Table 5.6, at a stress of σ = 90 MPa the rate of migration in the systems Ni(Cu) and Ni(Ag) is 1.5–2 times smaller in comparison with nickel. However, the increase of applied stress increases the value of V in nickel, and in the system Ni(Cu) and Ni(Ag) this parameter decreases. This decrease of the rate of migration with increasing load in these two systems is evidently associated with the following reasons: 1) with the appearance of stress gradients, suppressing grain boundary migration, at the grain boundaries because of the formation in the boundary regions of a solid solution [146]; 2) the formation and development of grain boundary cracks whose presence, as shown in [265], results in a decrease of the mobility of high-angle grain boundaries.

Figure 5.20a shows that in the system Ni(Ag) in the range of low strain rates at temperatures of 823 and 873 K the intensity of

Table 5.6. Dependence of the rate of migration of grains (V) in nickel and the Ni(Cu) and Ni(Ag) systems on applied stress in creep (T = 873 K)

Material	V, m/s		
	σ = 90 MPa	σ = 110 MPa	σ = 150 MPa
Ni	$2.4 \cdot 10^{-10}$	$3.0 \cdot 10^{-10}$	$3.9 \cdot 10^{-10}$
Ni (Cu)	$3.9 \cdot 10^{-10}$	$3.6 \cdot 10^{-10}$	$2.6 \cdot 10^{-10}$
Ni (Ag)	$6.1 \cdot 10^{-10}$	$1.2 \cdot 10^{-10}$	$8.0 \cdot 10^{-11}$

the effect of activation of creep by grain boundary diffusion flows decreases. Evidently, this is caused by the high mobility of the grain boundaries at low creep rates (low applied stresses). As shown in [81], during migration, the grain boundary may transfer from the diffusion regime B1 to regime C which corresponds to the initial filling of the grain boundary with the solute. This may be the reason for a decrease or complete disappearance of the effect of acceleration of creep. This claim is confirmed by the marked deviation of the values of V at T = 873 K in the rate of low creep rates ($6.1 \cdot 10^{-10}$ cm/s) from the values calculated for the Ni(Ag) system using the model proposed in [144] ($1.1 \cdot 10^{-10}$ cm/s). The rate of migration is up to six times higher than the maximum permitted rate on the basis of the conditions of maintaining the regime of type B1 [81].

As shown previously, the migration of the grain boundaries in the process of diffusion results in the saturation of the volume behind the migration front with the solute atoms. The saturation results in solid solution hardening of the material – the base as a result of which the creep rate of the polycrystal, subjected to the effect of grain boundary diffusion flows of the solute, may greatly decrease in comparison with the pure material.

Thus, the grain boundary diffusion flows of the solute atoms from the surface in the conditions of non-stationary diffusion B1 cause activation of sliding through high-energy grain boundaries in nickel polycrystals and, consequently, a decrease of the creep resistance.

The results show that the increase of the length of the grain boundaries and of the degree of diffusion permeability of the grain boundaries should result in a large increase of the effect of acceleration of creep in the conditions of the effect of grain-boundary diffusion flows of the solute atoms from the surface and the shift of the maximum of manifestation of the effect in the direction of low temperatures. From this viewpoint, the physical-chemical and mechanical properties of nanostructured materials, produced by the effect of severe plastic deformation and characterised by considerable nonequilibrium and high diffusion mobility of the grain boundaries, may have a low resistance to the effect of the external medium, associated with the diffusion of the solute along the grain boundaries, already at temperatures of 0.2–0.3 T_m.

Below, we discuss the results of investigations carried out by the authors of [169, 170, 249, 250, 266] into the mechanical properties of the nanostructured materials in creep in the conditions of the effect of the diffusion flows of the solute from the surface. The materials for the investigations were the Ni(Cu) and Cu(Al) systems, and the nanostructures in the material – base were produced by severe plastic deformation. The creep test was carried out on the materials in the condition immediately both after the effect of severe plastic deformation and after pre-recrystallisation annealing.

The calculations of the temperature range of existence of the diffusion regime of type B1 in the investigated nanostructured materials were carried out on the basis of the modified Fisher model [66, 147] (this calculation procedure is justified in chapter 3). According to this model, the diffusion conditions in the grain boundaries of the nanostructured materials are similar to those considered in the Fisher model [81], with the only difference being that in the equations for the asymptotic parameters and the coefficients of volume diffusion are replaced by the product $g^2D_d = D_{ot}$ (where g is the volume fraction of the defects, D_d is the coefficient of diffusion through defects). For the known maximum density of the dislocations in the nanostructured nickel and copper of respectively 7·1014 m^{-2} [202] and $5 \cdot 10^{14}$ m^{-2} [285] the maximum possible values of D_{ot} are 6–7 orders of magnitude higher than the values of the appropriate coefficients of volume diffusion in coarse-grained nickel and copper. Substituting the resultant values of D_{ot} into the equation for λ and ν, it may be seen that the realisation of the conditions, corresponding to the diffusion regime B_1, in the nanostructured nickel and copper, corresponds to the temperature range 398–473 K. Consequently, the maximum of the effect of acceleration of creep for the nanostructured nickel and copper, estimated from the ratio $\dot{\varepsilon}_1/\dot{\varepsilon}_2$ ($\dot{\varepsilon}_1$ and $\dot{\varepsilon}_2$ are the rates in the steady-stage creep in vacuum and in the conditions of diffusion of the solute

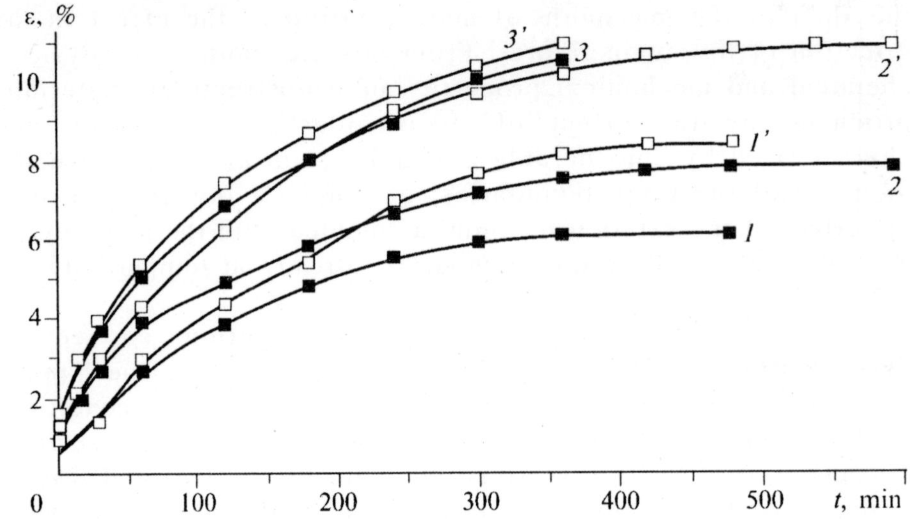

Fig. 5.23. Creep curves of the specimens of nanostructured nickel at temperatures of 398 (*1*, *1'*), 423 (*2*, *2'*) and 523 (*3*, *3'*) in vacuum (*1*–*3*) and in the presence of grain boundary diffusion flows of copper (*1'*–*3'*); σ = 630 MPa.

atoms from the surface, respectively) should be detected in the given temperature range.

In fact, as shown in Figs. 5.23–5.26, the effect of acceleration of the creep of these nanostructured materials in the presence of the diffusion flows of the solute atoms from the surface is detected in the temperature range 398–473 K, i.e. at considerably lower temperatures in comparison with the corresponding temperatures for coarse-grained nickel (773–923 K [249]) and copper (573–673 K [240]).

It should be mentioned that the nature of creep of nanostructured nickel in the condition immediately after severe plastic deformation in vacuum and in the conditions of the effect of diffusion flows of the impurity atoms from the surface is the same and corresponds to logarithmic creep (Fig. 5.23). However, the creep rate in the non-steady creep of nanostructured nickel in this condition under the effect of the diffusion flows of the impurity atoms is higher than in creep in vacuum. This is also accompanied by an increase of the degree of plastic deformation at which the creep curves reach saturation.

Thus, regardless of the fact that the grain boundaries in nanostructured nickel, produced by the effect of severe plastic the formation, are highly nonequilibrium, the effect of the diffusion flows of the copper atoms, as in the coarse-grained material,

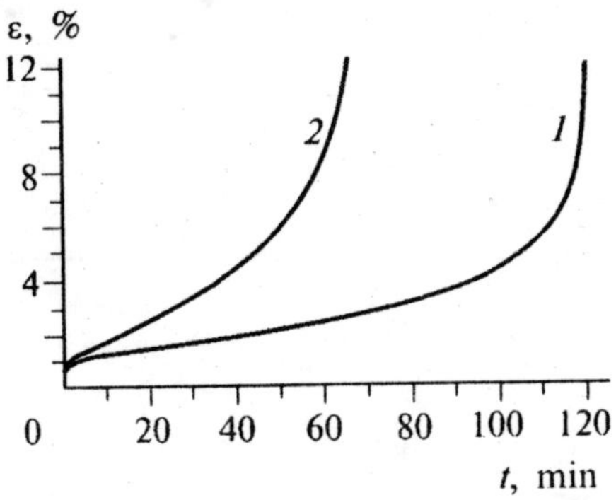

Fig. 5.24. Creep curves of specimens of nanostructured copper in vacuum (*1*) and in the conditions of diffusional aluminium from the surface (*2*); T_{test} = 423 K, σ = 250 MPa.

results in their activation, manifested in a decrease of the creep resistance.

A more detailed examination of the effect of diffusion flows of the solute atoms from the surface was carried out by the authors on the example of nanostructured nickel, subjected to the effect of severe plastic deformation by the method of equal-channel angular pressing with annealing at T = 398 K for 1 h. Figure 5.26 shows that in the nanostructured nickel, as in the coarse-grained material, the curves of the dependence of the strain rate on stress are linear with the exponent at the stress index n of 9.

In the conditions of grain boundary diffusion of the copper atoms from an external source, the identical curves for nanostructured nickel in the entire examined temperature range are characterised

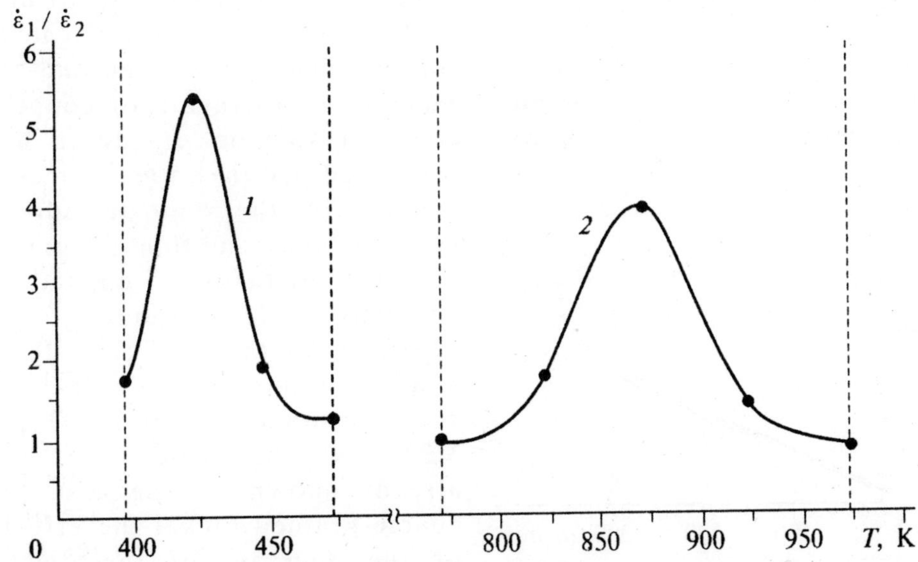

Fig. 5.25. Temperature dependence of the effect of activation of creep by grain boundary diffusion flows of copper atoms for nanostructured (*1*) and coarse-grained (*2*) nickel. The dotted line indicates the regions in which diffusion of copper takes place in regime B_1. The strength of the effect is characterised by the ratio of the creep rate in the conditions of the effect of diffusion flows through the effect in a vacuum.

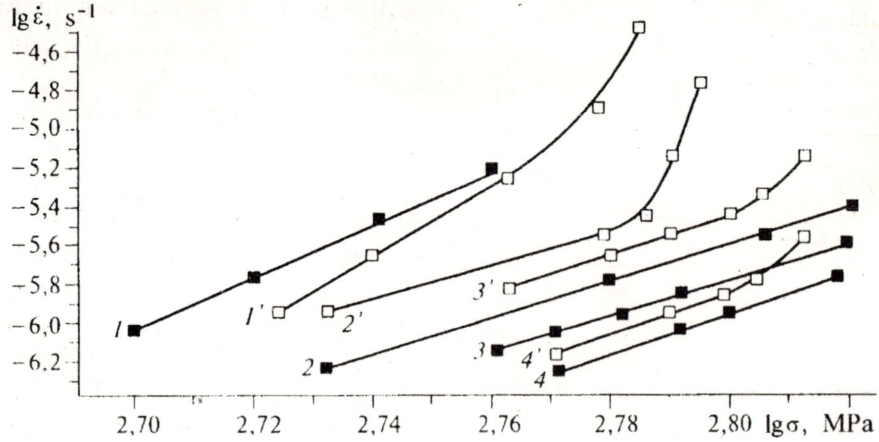

Fig. 5.26. Dependence of the steady-state creep rate of applied stress for nanostructured nickel in vacuum (*1–4*) and in the conditions of diffusion of copper from the surface (*1'–4'*) at temperatures of 473 (*1, 1'*), 443 (*2, 2'*), 423 (*3, 3'*) and 398 (*4, 4'*).

by a distinctive two-stage form (Fig. 5.26). In the region of low stresses at temperatures of 398, 423 and 444 K, the angle of inclination of the investigated curves coincides with the appropriate angle for pure nanostructured nickel, whereas at high stresses the exponent at the stress index n increases rapidly to 30–50. At all temperatures, with the exception of T = 473 K, in the investigated applied stress range the effect of the increase of the creep rate of nanostructured nickel in the conditions of diffusion of copper increases in comparison with creep in vacuum. It should be mentioned that the nature of these curves for the nanostructured nickel in the conditions of grain boundary diffusion of copper is qualitatively identical to the curve obtained for the Ni(Cu) system at a temperature of 773 K for the coarse-grained material (Fig. 5.21). At the same time, as already shown, in the case of coarse-grained nickel the effect of the decrease of the creep resistance under the effect of diffusion flows of the solute is associated with the activation of grain boundaries and, conseq-

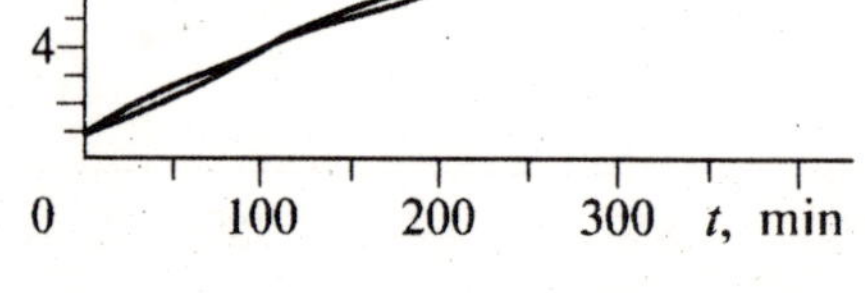

Fig. 5.27. Creep curves of the specimens of Cu–0.5 wt% Al_2O_3 nanostructured composite in vacuum (*1*) and in the conditions of diffusion of aluminium from the surface (2); T_{test} = 423 K, σ = 285 MPa.

Table 5.7. Apparent activation energy of creep in nanostructured nickel and Ni(Cu) system

T, K	$Q_c \pm 15$ kJ/mol	
	Ni	Ni (Cu)
423	114	70
443	112	71
473	110	108

uently, an increase of the rate and contribution of grain boundary sliding controlled by grain-boundary diffusion, to the total deformation. On the basis of these considerations, it may be assumed that the given effect for the nanostructured materials is of the same nature.

The measurement of the apparent activation energy of creep (Q_c) of nanostructured nickel, carried out by the method of the temperature jump, confirms this assumption. As indicated by Table 5.7, the value of Q_c of the nanostructured nickel in creep in the conditions of diffusion of copper atoms from the surface in the temperature range characterised by the effect of acceleration of creep, decreases to ~70 kJ/mole, which is close to the activation energy of grain boundary diffusion of copper in nanostructured nickel (~60 kJ/mole [250]). At the same time, Q_c of nanostructured nickel at a temperature of 473 where the creep resistance increases at low strain rates, is almost equal to the appropriate value of Q_c in creep in vacuum.

It should be mentioned that the effect of the increase of the creep resistance under the effect of the diffusion flows of the solute atoms of nickel was previously detected in polycrystalline iron [134, 286]. It was shown in [134] that this behaviour is caused by the variation of the conditions of grain-boundary diffusion from B_1 to B_2. This results in the saturation of the boundary regions of iron with nickel atoms preventing the movement of dislocation. It may be assumed that the identical situation is also detected in nanostructured nickel in creep in the conditions of diffusion flows of copper at a temperature of 473 K.

Examination of the structure of nanostructured nickel, carried out by transmission electron microscopy, confirmed this assumption. For example, in the specimens of nanostructured nickel after creep in the conditions of copper diffusion at a temperature of 473 K at a depth of 10 μm (in the diffusion zone of penetration of copper atoms from the grain boundaries of nickel) examination showed a large increase of the dislocation density in comparison with the specimens investigated after creep at a lower temperature (443 K). Evidently, this increase is the result of saturation of the boundary regions of nickel with copper atoms, blocking the departure

Table 5.8. Effect of atmosphere on the creep resistance of copper and the Cu-0.5 wt% Al_2O_3 composite (test temperature 673 K).

Material		Vacuum		Air	
		σ, MPa	ε_1, s^{-1}	σ, MPa	ε_2, s^{-1}
Copper (recrystallised)		31	$9.3 \cdot 10^{-6}$	31	$9.8 \cdot 10^{-7}$
Composite, condition	1	70	$4.6 \cdot 10^{-7}$	70	$3.1 \cdot 10^{-8}$
	2	75	$3.8 \cdot 10^{-6}$	160	$4.7 \cdot 10^{-8}$
				190	$1.1 \cdot 10^{-7}$
	3	75	$2.8 \cdot 10^{-7}$	75	$5.3 \cdot 10^{-8}$

of dislocations to the sinks (grain boundaries). This also results in the increase of the creep resistance of nanostructured nickel in the conditions of copper diffusion observed at a high temperature (473 K).

Composites

Examination of the effect of diffusion flows of the solute atoms from the external medium on the creep of ultrafine great composites was carried out by the authors of [187, 240] on the example of Cu-0.5 wt% Al_2O_3 composites. The nanostructure of this composite in a copper die was produced by the effect of severe plastic deformation by torsion under pressure in equipment identical to the Bridgman chamber. The diffusant was represented by aluminium.

As indicated by Fig. 5.27, the creep curves of the Cu–0.5 wt% Al_2O_3 nanostructured composite in vacuum and in the conditions of diffusion of the solute from the surface (coating) are almost identical. At the same time, for the pure nanostructured copper in the same testing conditions, the creep rate increases approximately 5 times as a result of the effect of the diffusion flows of the aluminium atoms (Fig. 5.24).

It was shown previously that the effect of activation of creep of polycrystals by grain-boundary diffusion flows of the solute atoms is associated with the activation of grain boundary sliding and with the increase of the contribution of this process to total strain. It may be assumed that the absence of this effect in the creep of the

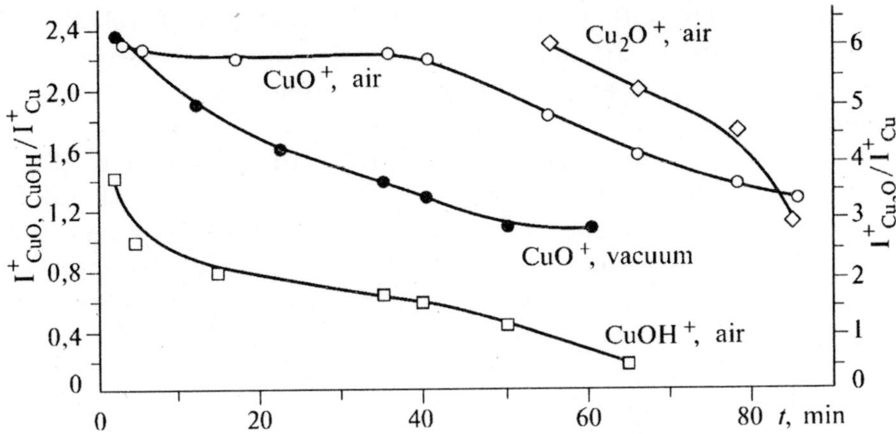

Fig. 5. 28. Dependence of the intensity of emission of secondary ions on the sputtering time of the surface of specimens after creep in vacuum and in air (HRMS data).

nanostructured Cu–0.5 wt% Al_2O_3 composite in the conditions of diffusion of the aluminium atoms from the surface is caused by the well-known effect of the suppression of grain boundary sliding by the dispersed particles, distributed at the grain boundaries [287].

Interesting results were obtained by the authors in examining the effect of atmosphere on the creep of ultrafine grained composite Cu–0.5 wt% Al_2O_3. Tests were carried out on the specimens of the composite in three conditions: after hot extrusion (condition 1), extrusion and subsequent deformation by rolling by 70% (condition 2), extrusion and subsequent deformation by torsion (condition 3 – nanostructured composite), and also pure recrystallised copper. The structure of the composite after these treatments is shown in section 4.1.

Table 5.9. Effect of the thickness of the specimen on the creep resistance of copper and the Cu–0.5 wt% Al_2O_3 composite in air (test temperature 673 K)

Material		Creep rate, s^{-1}	
		Thick specimen	Thin specimen
Copper (recrystallised)		$9.8 \cdot 10^{-7}$	$3.6 \cdot 10^{-7}$
Composite, condition	1	$3.1 \cdot 10^{-8}$	$3.1 \cdot 10^{-8}$
	2	$4.7 \cdot 10^{-8}$	$9.0 \cdot 10^{-9}$
	3	$5.3 \cdot 10^{-8}$	$2.1 \cdot 10^{-8}$

Table 5.8 shows that the creep resistance of the Cu–0.5 wt% Al_2O_3 composite in all investigated conditions, and also of pure copper, in testing in air is considerably higher than in testing in vacuum.

Analysis of the elemental composition of the surface of the specimens of the Cu–0.5 wt% Al_2O_3 composites, carried out by high-resolution Mössbauer spectroscopy (HRMS), shows that after creep in air the surface layers of the material are characterised by increased concentration of the copper oxides in comparison with the concentration of the deep layers (Fig. 5.28). With increase of the distance from the surface, the concentration of the copper oxides decreases.

On the basis of these results it may assumed that the increase of the creep resistance of the Cu–0.5 wt% Al_2O_3 copper and pure copper in testing in air is associated with the saturation of the surface layer of the material with oxygen which leads to solid solution and dispersion hardening. In order to verify this assumption, creep tests were carried out on specimens of pure coarse-grained copper and the Cu–0.5 wt% Al_2O_3 composite in the conditions 1–3 of two different thicknesses (the thick specimens were approximately twice thicker than the thin specimens). The test results show (Table 5.9) that the thin specimens of pure coarse-grained copper and the Cu–0.5 wt% Al_2O_3 composite in all investigated conditions have higher creep resistance than the thicker specimens. This confirms the previously made assumption on the nature of the observed difference of the creep rates in vacuum and in air.

CHAPTER 6

EFFECTS OF THE STRUCTURAL-PHASE CONDITION ON ELASTOPLASTIC PROPERTIES OF NANOSTRUCTURED TITANIUM IN STATIC AND CYCLIC LOADING

As mentioned in the previous chapters, the typical feature of nanostructured metals, produced by the method of equal-channel angular pressing, in addition to the ultrafine grains, are high-angle grain boundaries with an imperfect structure, i.e. elastic strains and higher long-range stresses, and increased dislocation density in comparison with recrystallised polycrystalline metals [150]. The ultrafine structures, produced by this method, are typical by high stored energy and, consequently, low thermal stability.

As shown in chapter 4, the annealing of metals with this structure is accompanied by the redistribution of dislocations and a decrease of the density of dislocations inside the grains, structural recovery of non-equilibrium boundaries with a reduction of the level of long-range stress fields and then recrystallisation and growth of grains in the recrystallised structure [150]. Since the activation energy of diffusion of interstitial impurities is approximately 50% of the activation energy of self diffusion [197, 288], the evolution of the dislocation substructure in annealing of these metals may be accompanied by the redistribution of interstitial impurities. Because of the presence of gradients of long-range stress fields at these grain boundaries, all diffusion processes around these boundaries should be accelerated and this should change the deformation behaviour in the microdeformation region and in fatigue loading.

Here, we present the results of investigations of the changes in elastic-plastic properties in the microdeformation region in bending, torsion and fatigue loading, caused by plastic deformation and annealing of nanostructured titanium produced from coarse-grained titanium VT1–0 by the method of equal-channel annular pressing.

6.1. MICROPLASTIC DEFORMATION OF NANOSTRUCTURED TITANIUM IN STATIC AND CYCLIC LOADING

Nanostructured materials are structurally heterogeneous media with a set of different orientations, form and grain size. Theoretical studies [289–291] show that because of these special features of the grain structure and anisotropy of the elastic moduli, the initial stage of loading of the coarse-grained and nanostructured metals results in the presence of stresses of different magnitude in the grains. This heterogeneity of the stress state of the metal and the presence of grain boundary stress microconcentrators with different efficiency should result in gradual inclusion of grains in plastic deformation. This has been confirmed by experiments for coarse-grained polycrystals [291]. Regardless of the phase and structural states of the polycrystal, plastic deformation initially takes place in individual grains which are not in contact with each other on the surface of the specimen [291–294]. In each grain, sliding takes place in a specific system; slip bands start at the boundary and gradually extend into the body of the grains. The stress σ', at which the first signs of sliding are detected, is independent of the grain size [291, 293]. This stress is regarded as a true limit of elasticity of the polycrystal.

The increase of stress increases the number of surface plastically deformed grains. Finally, at a specific external stress $\sigma = \sigma''$, one or several groups of plastically deformed grains situated next to each other appear, i.e. the co-operative plastic deformation of grains starts to take place. Stress σ'' is regarded as the macroscopic limit of elasticity of the polycrystal [291, 293]. Now, the increase of the applied stress results in the activation of dislocation sources at the grain boundaries whose distance from the surface continuously increases. In plastically deformed metals, there are already two or more slip systems.

The start of co-operative plastic deformation of the grains denotes the change of the mechanism of combined deformation of the plastically deformed grains with the adjacent grains with elastic accommodation in plastic shear. From this viewpoint, the macroscopic limit of elasticity is the critical characteristic of the polycrystal and has a fully defined physical meaning. In contrast to the true elasticity limit, this limit depends on the grain size according to the Hall–Petch equation [291].

The appearance of a group of plastically deformed grains at $\sigma = \sigma''$ indicates the nucleation of a Lüders–Chernov band [291]. In further stages, plastic deformation develops by the growth of these bands. In addition to the grain size, the number of these bands also depends on the phase and structural state of the polycrystal (mainly on the difference between the start stress and the resistance to movement of dislocations in the grains) [291]. As the difference increases, the number of nucleated and formed Lüders–Chernov bands decreases.

The stress above which the formation of the first Lüders-Chernov band is completed is also a critical characteristic of the polycrystal (it is regarded as the physical yield limit) [291]. At this stress, all the grains in one of the sections of the specimen are initially involved in plastic deformation. On the yielding curve, the physical yield limit corresponds to the upper yield limit, and in its absence, to the start of the yield plateau. It characterises the end of the region of microplastic deformation [291]. On the yield plateau, plastic deformation develops by the propagation of the Lüders–Chernov bands.

The relationships of buildup of microplastic strain were explained after developing high-accuracy methods of stress measurement. It has been established that in polycrystalline metals and alloys, microplastic strain ε_{pl} builds up with the increase of applied stress in two stages (Fig. 6.1) [291]. In the first stage, the relationship between the applied stress and the degree of plastic deformation is linear, in the second stage it is parabolic. Here, the first stage corresponds to plastic deformation of grains isolated from each other, the second one to the co-operative plastic deformation of the grains. The transition from the first to second stage of the yielding curve is associated with the start of co-operative plastic deformation of the grains, i.e. corresponds to the macroscopic elasticity limit σ''.

If an alloy contains particles of a second phase and the contribution of these particles to the resistance to movement of dislocations is greater than the contribution of the atoms of alloying elements and interstitial impurities, at the stresses below the macroscopic yield limit σ'' the curve $\sigma(\varepsilon)$ shows two sections of

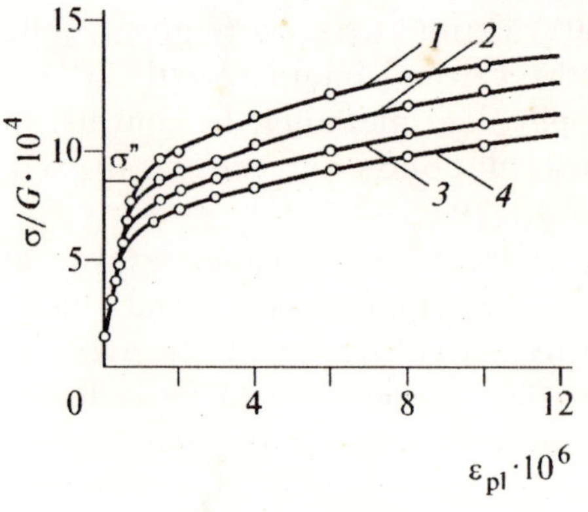

Fig. 6.1. Loading diagram (torsional deformation) of polycrystals Cu + 9.2 at% Al, T = 163 K. 1) d (grain diameter) = 15 μm, 2) 25, 3) 47, 4) 90 μm.

the linear dependence of ε_{pl} on σ [291].

The relationships governing the buildup of microplastic deformation in quasi-static loading of ultrafine grains materials have been examined only for nanostructured titanium. In [295] it was established that in nanostructured titanium, as in coarse-grained titanium, the physical yield limit is preceded by microplastic deformation. The change of the coarse-grained structure by the nanostructure has no effect on the occurrence of individual stages and relationships of the buildup of microplastic deformation ε_{pl} (Fig. 6.2). In both structural conditions the curves $\sigma(\varepsilon_{pl})$ show below the yield limit two sections of linear and one section of parabolic relationship between σ and ε_{pl}.

The published experimental data taking into account the investigations [291] of polycrystalline materials indicate that the linear sections of the curves $\sigma(\varepsilon_{pl})$ in coarse-grained and nanostructured titanium correspond to the first stage of microplastic

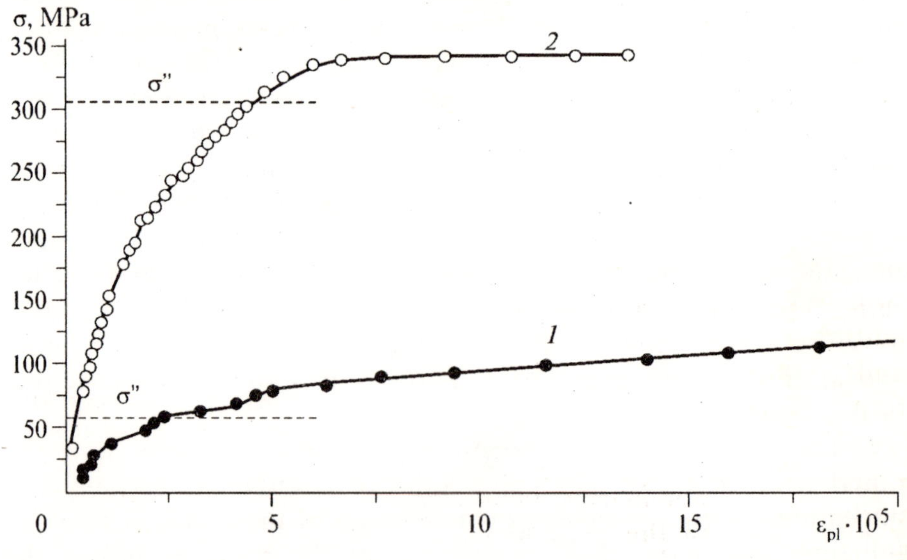

Fig. 6.2. Buildup of microplastic strain in torsion for coarse-grained (*1*) and nanostructured (*2*) titanium.

deformation in which plastic deformation takes place in individual grains which are not in contact with each other; the second stage of macroplastic deformation shows the start of parabolic strengthening.

The presence of two sections on the curves $\sigma(\varepsilon_{pl})$ in the first stage of microplastic deformation is determined by the heterophase condition of titanium in both the coarse-grained and nanostructured states. The stress σ'' of transition from the first stage of microplastic deformation to the second stage in nanostructured titanium is considerably higher than in coarse-grained titanium; the physical limit also increases several times [198, 295].

The variation of the parameters of the nanostructure in pre-recrystallisation annealing does not change the relationships of the buildup of microplastic deformation [198, 295]. There are only changes in the characteristics of microplastic deformation: the strengthening coefficients in the first and second stage of microplastic deformation and the macroscopic yield limit.

As in the case of coarse-grained titanium, the mechanical properties of nanostructured titanium depend on its structural–phase condition [198]. The blocking of dislocations, dislocation substructure, the condition of the grain boundaries and the nature of distribution of the interstitial impurities in the body of the grains and at the boundaries of nanostructured titanium can be changed in severe plastic deformation and subsequent pre-recrystallisation annealing. In addition to this, there may also be changes in the long-range internal stresses and in the capacity of the grain boundaries to emit and absorb lattice dislocations.

The effect of cold plastic deformation and subsequent annealing on the deformation behaviour of nanostructured titanium in the range of microplastic deformation was investigated in [198, 295]. The main result of these investigations may be described as follows. The nature of buildup of microplastic deformation with increasing stress is independent of the method of production and subsequent treatment of nanostructured titanium, i.e. of the structural–phase condition; the two-stage buildup of microplastic deformation is still found (Fig. 6.3). However, the characteristics of microplastic deformation (the coefficients of strengthening in the first and second stage of microplastic deformation, the macroscopic yield limit and the yield limit) depend on the structural-phase condition of the material. Table 6.1 gives the values of the macroscopic yield limit for different structural–phase conditions. For comparison, the table also gives the data on the mechanical properties and microhardness.

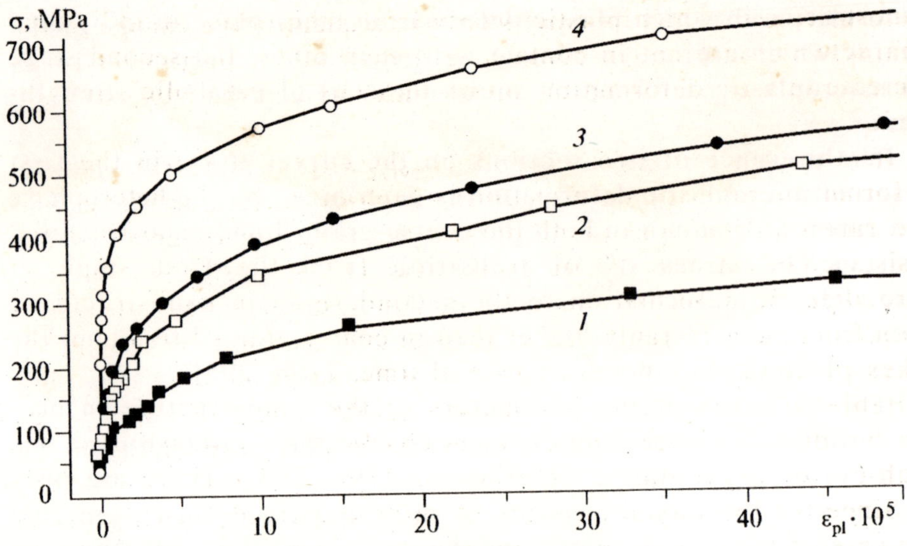

Fig. 6.3 Buildup of microplastic deformation in bending coarse-grained (*1*) and nanostructured titanium after equal-channel angular pressing (*2*), subsequent cold deformation (*3*) and pre-recrystallisation annealing (*4*).

Table 6.1. Mechanical properties of titanium

Material	σ''	σ_B	$\sigma_{0.2}$	H_μ	Δ, %
	MPa				
Coarse-grained	180	464	360	1630	24.8
Nanostructured	260	716	654	2300	18.9
Cold-worked	220	1000	947	2780	8.8
Annealed	380	960	860	2950	10.6

For example, cold plastic deformation reduces the macroscopic yield limit in comparison with the initial condition. At the same time, subsequent pre-recrystallisation annealing at a temperature of 523–573 K increases this limit several times with a simultaneous decrease of the rate of buildup of microplastic strain.

On the basis of these results it may be concluded that the main factor which determines the relationships of buildup of plastic deformation in the initial stage of loading in nanostructured material is, as in the case of coarse-grained materials, the heterogeneity of the stress states. The transition from the coarse-grained to nanostructured condition results only in a quantitative change of the characteristics of microplastic deformation, in particular in an increase of the macroscopic limit of elasticity and the physical yield limit.

In the cyclic loading of polycrystalline material, plastic deformation takes place even in the high-cycle fatigue range, and the rate of development of plastic deformation determines the resistance of the material the cyclic loading [291, 296–299]. In non-zero alternating loading the loops of the elastoplastic hysteresis are open from the very beginning of plastic deformation, i.e. cyclic creep takes place [291, 299]. Therefore, this loading method is highly suitable for examining the relationships and mechanisms of the buildup of residual strain in cyclic loading, especially in the high-cycle fatigue range.

Since the endurance limit in the nanostructured materials, as in the case of the coarse-grained materials, corresponds to the region of microplastic deformation, the cyclic creep of titanium in the nanostructured and polycrystalline conditions has been investigated at the maximum cycle stress below the physical yield limit [198, 295]. Titanium was investigated in the same phase-structural condition as in the examination of microplastic deformation under quasi-static loading. The specimens were subjected to cyclic loading at room temperature with a frequency of 1.7 Hz in uniform bending of a flat specimen.

These investigations gave the following results [198, 295]. In the nanostructured and coarse-grained titanium, residual deformation in cyclic loading builds up with the increase of the number of cycles in accordance with the same law, i.e. the logarithmic law which is violated only prior to fracture (Fig. 6.4). In nanostructured titanium, as in coarse-grained titanium and other polycrystalline materials [291], the rate of buildup of residual strain and its magnitude depend on the maximum cycle stress (Figs. 6.4, 6.5). If the value of the latter does not exceed the macroscopic yield limit σ'', a small residual strain of the same order as in quasi-static loading builds up. When the macroscopic limit of elasticity is greatly exceeded, the rate of buildup of residual strain rapidly increases.

For the same number of cycles in loading the dependence of residual strain ε_{pl} on the maximum cycle stress σ_{max} for the nanostructured and coarse-grained titanium is qualitatively identical to that in quasi-static loading (Fig. 6.6). The first stage of microplastic deformation on the curves ε_{pl} (σ_{max}) corresponds to two sections of the linear dependence of ε_{pl} on σ_{max}, and parabolic

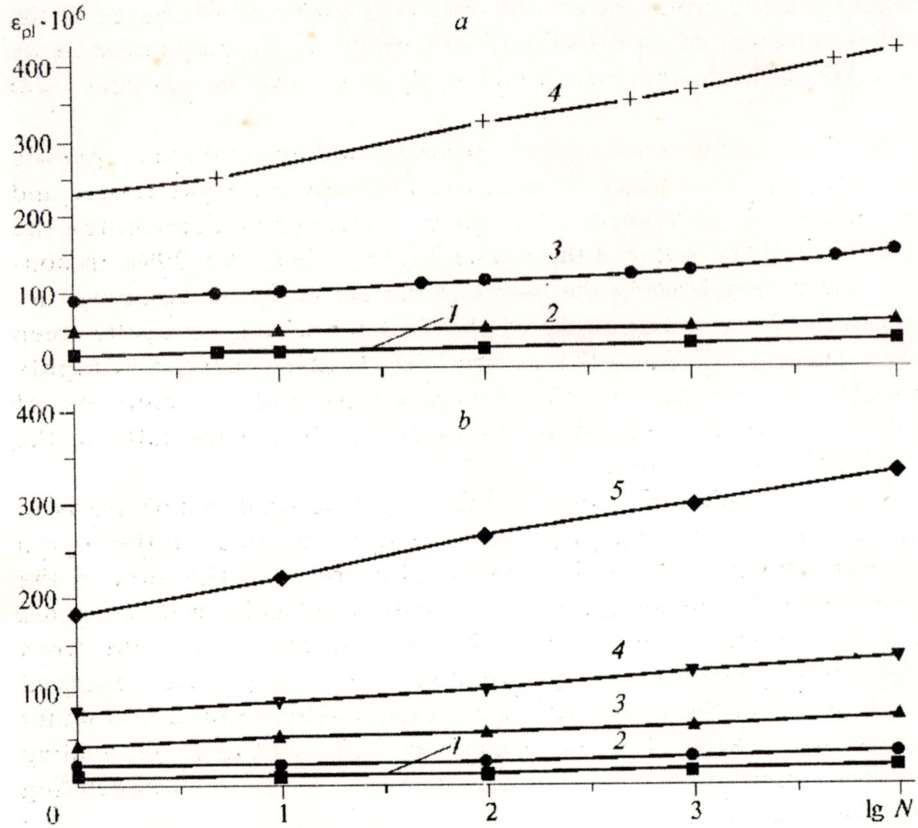

Fig. 6.4. Buildup of microplastic strain in the cyclic loading of coarse-grained (*a*) and nanostructured (*b*) titanium. Maximum cycle stress. *a*: *1*) 120, *2*) 140, *3*) 205, *4*) 290 MPa; *b*: *1*) 175, *2*) 230, *3*) 290, *4*) 350, *5*) 430 MPa.

strengthening is recorded in the second stage. The maximum cycle stress at which the linear strengthening is replaced by parabolic one, is approximately equal to the macroscopic elasticity limit. A similar agreement between the stage of buildup of plastic deformation in quasi-static and cyclic loading was recorded in [299] for polycrystalline alloys.

As already mentioned, the residual deformation after the same number of cycles at different values of the initial stress amplitude for nanostructured titanium (Fig. 6.6, *2*) is always lower in comparison with coarse-grained titanium (see Fig. 6.6, *5*). High-intensity plastic deformation has almost no influence on this value. At the same time, pre-recrystallisation annealing resulted in a rapid decrease of residual strain (Fig. 6.6, *4* and *6*).

The results of fatigue tests for titanium in different conditions

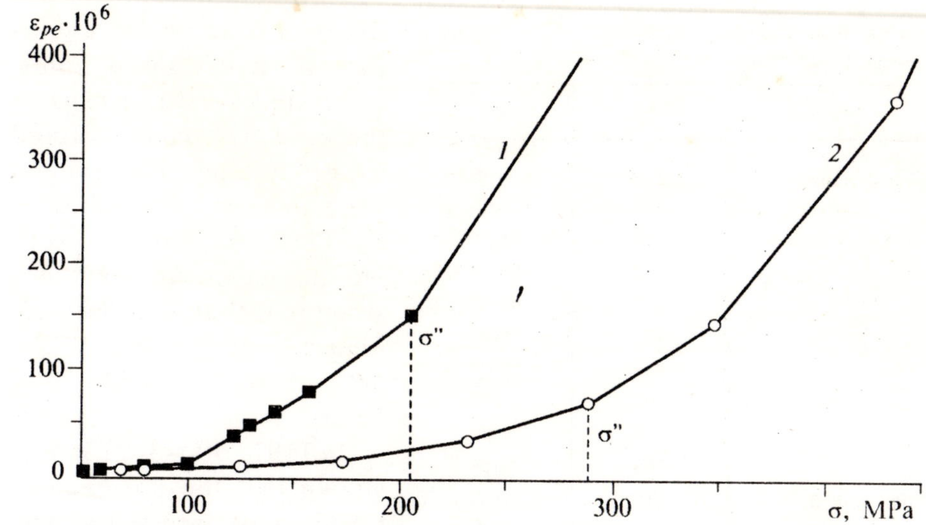

Fig. 6.5. Dependence of residual strain on the maximum cycle stress at 10^4 cycles. *1*) coarse-grained, *2*) nanostructured titanium.

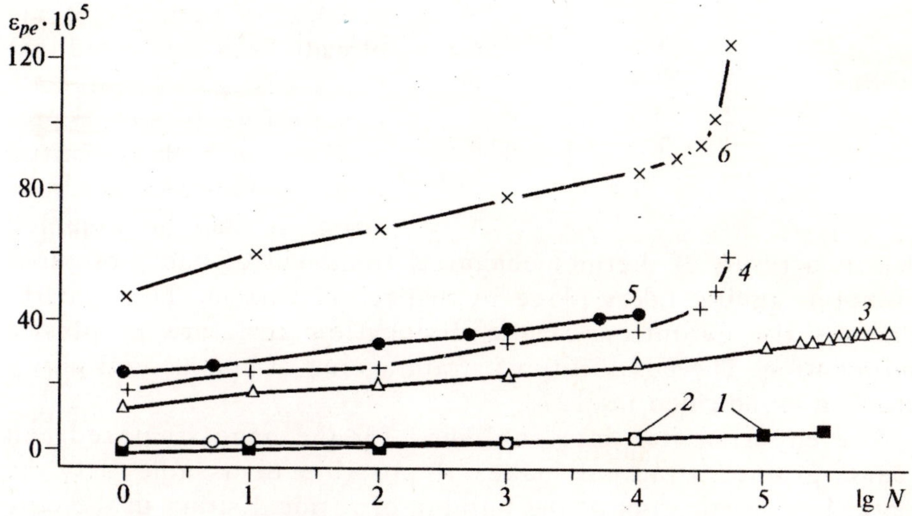

Fig. 6.6. Buildup of microplastic strain in cyclic loading in relation to the number of cycles for coarse-grained titanium (*1*, *5*), nanostructured titanium in the initial (*2*) condition, after cold deformation (*3*, *6*), after pre-recrystallisation annealing (*4*). Maximum cycle stress, MPa: *1*) 122, *2*) 260, *3*) 417, *4*) 625, *5*) 288, *6*) 625.

and at different maximum cycle stresses are presented in Table 6.2. For all conditions at the maximum cycle stresses σ_{max}, close to the macroscopic elasticity limit σ", the specimens did not fail even after 6 load cycles. At the same time, the endurance limit for finite

Table 6.2. Fatigue characteristics of the investigated materials

Material	σ_{max}, MPa	*N*
Coarse-grained	122	$>10^6$
	351	$>10^6$
	351	$40\cdot10^6$
	362	$40\cdot10^4$
	375	$42\cdot10^4$
	430	$2\cdot10^4$
	471	$25\cdot10^4$
Nanostructured	260	$>10^6$
	509	$>10^6$
	598	$2,10^4$
Cold-worked	417	$>10^6$
	520	$>10^6$
	534	$18.5\cdot10^4$
	552	$9\cdot10^4$
	562	$12\cdot10^4$
	573	$25\cdot10^4$
	625	$6\cdot10^4$
Annealed	469	$>2.1\cdot10^6$
	520	$>10^6$
	625	$6\cdot10^4$

life at a base of 10^6 cycles for all investigated nanostructured conditions, as in the case of coarse-grained titanium, does not exceed yield limit $\sigma_{0.2}$. As indicated by Table 6.2, at non-zero alternating loading of nanostructured titanium, the endurance limit for finite life at 10^6 cycles exceeds 500 MPa.

In [197, 300–302] it was shown that the presence in titanium of interstitial impurities (oxygen, nitrogen, carbon) may greatly affect the mechanical characteristics of titanium. It has already been mentioned that the investigated materials contain fine-dispersion precipitates in both the initial and nanostructured condition. It may be assumed that as a result of thermomechanical treatment of nanostructured titanium, ageing takes place in respect of carbon. This greatly changes the evolution of the dislocation structure in plastic deformation. The possibility of strain ageing of commercial purity titanium is indicated in [197].

The experimental data published for the nanostructured and coarse-grained titanium makes it possible to assume that the marked intensification of the buildup of residual strain in the case of high applied stresses above the macroscopic yield limit is caused by the change in the mechanism of combined deformation of the grains with elastic accommodation for plastic shearing. This indicates [291] that, as in the case of the coarse-grained polycrystals, the endurance limit of nanostructured titanium for non-zero alternating loading is approximately equal to the macroscopic elasticity limit σ''. Cyclic tests over 10^6 cycles show that at the maximum cycle stress lower than the macroscopic elasticity limit the specimens of nanostructured and coarse-grained titanium did not fracture and did not contain micro- and macrocracks [198, 295].

6.2. EFFECT OF ULTRASOUND DEFORMATION OF THE SURFACE ON THE STRUCTURE AND MECHANICAL PROPERTIES OF COARSE-GRAINED AND NANOSTRUCTURED TITANIUM

Many properties of metals and alloys are determined to a large degree by the structural–phase condition of surface layers. Therefore, for the efficient variation of both physical-chemical and mechanical properties, extensive use is made of special methods of mechanothermal and chemical effects resulting in changes in the structural-phase condition only in the surface layers [303–305]. With other conditions being equal, the nature of the effect of any of these methods on the structure of the surface layers (and, consequently, on the properties) depends on the initial condition of metal (on the density and distribution of dislocations, grain size, etc).

The nanostructured metals, produced by the effect of severe plastic deformation, in particular by the method of equal-channel angular pressing, in contrast to recrystallised coarse-grained metals, are characterised by high stored elastic energy which is closely linked with the nonequilibrium grain boundaries [198, 295, 306, 307]. Consequently, for the same method of the effect, the structural-phase condition in the surface layers in the metal with a nanostructure may differ from that of coarse-grained metal, i.e. there may also be differences in the changes of the physical-chemical and mechanical properties. Therefore, the data on the variation of the properties in surface treatment of coarse-grained metals cannot be transferred in a general case to nanostructured metals. Therefore, it is interesting to examine the data [308] on the effect of surface quasi-plastic deformation on the structure and mechanical behaviour in the static and cyclic loading of coarse-grained and nanostructured titanium, produced by the effect of severe plastic deformation.

In [308], investigations were carried out on specimens of titanium of commercial purity, grade VT1–0, in the initial recrystallised and nanostructured conditions. The nanostructure was produced by severe plastic deformation by the method of equal-channel angular pressing [150]. In the study, investigations were carried out into the effect of two types of ultrafine grain condition of titanium: the first – after equal-channel angular pressing, and the second one after equal-channel angular pressing and subsequent volume severe plastic deformation. Surface plastic deformation was carried out by ultrasonic deformation.

Ultrasonic deformation of the surface of the materials carried out to increase the service characteristics is well-known [309–314] and is used quite widely in industrial practice. Ultrasonic treatment of the surface of the specimen was carried out using equipment Moskit [315]. The tool used in this equipment was a rod made of a hard alloy of the VK type with a diameter of 5 mm with a spherical tip. The generator of ultrasound oscillations excited oscillations of ultrasound frequency in the rod. Consequently, the surface was deformed. Since Moskit equipment makes it possible to regulate the amplitude of ultrasound oscillations and stepped regulation of output power, the conditions of surface treatment can be varied in a wide range.

Flat specimens were used for the examination of the mechanical and fatigue properties. Therefore, a special method of ultrasound treatment of flat specimens with a thickness of 0.2–0.4 mm in Moskit equipment was developed. The specimen was placed on a specially produced platform with its long side parallel to the reciprocal movement of the tool. As a result of ultrasound treatment, the flat surfaces of the specimen on both sides were 'cross-hatched' in the longitudinal direction with deformation bands with a step of 0.2 mm.

The structure of the surface layers was examined by transmission electron microscopy in a EM–125K microscope. Microhardness was measured in equipment PMT–3M, the load on the indentor was 0.1 kg. The tensile test was carried out at room temperature at a strain rate of 10^{-3} s^{-1}. The loading-unloading method in the uniform bending of a flat specimen with a size of 46×6×(0.2–0.3) mm^3 using a former of the specific radius was used for measuring the microplastic deformation and determining the macroscopic elasticity limit σ'' [198, 295]. Fatigue tests were carried out in the same equipment in the non-zero alternating bending, frequency 1.7 Hz.

The structure of initial polycrystalline titanium and after equal-channel angular pressing is described in section 4.1. In ultrasound deformation by this method, the surface of the flat specimens of titanium in both the recrystallised and ultrafine grain conditions resulted in extensive distortion of the specimens. The middle of the specimen was always bent in the direction of the treated surface, indicating the formation of residual compressive stresses in the subsurface layer.

In ultrasound deformation, the most extensive changes in the microstructure of the subsurface layers took place in the specimens of coarse-grained recrystallised titanium. Dislocation density greatly increased, the material was divided into regions with the characteristic size of approximately 0.12 μm. Electron microscopic images of the boundaries between these regions are blurred and not sharp. Analysis of electron microdiffraction pattern shows that the spectrum of disorientation of the adjacent regions is in the range 2–40°. The microelectron patterns show broadening of the reflections and patterns are almost ring-shaped. The structure is similar to that corresponding to the metals subjected to the effect of severe plastic deformation with the formation of the nanostructured condition. These changes of the structure were observed in the subsurface layer of the investigated material with a thickness of approximately 100 μm.

In ultrasound deformation of the specimens of titanium in the nanostructured condition, the microstructure of the material is also slightly refined, and the characteristic dimensions of the structural elements are approximately 0.12 μm (Fig. 6.7*a*, *b*). The images of the microstructure of the specimens of deformed nanostructured titanium in the longitudinal section show that the grains remain elongated after ultrasound treatment but they are divided into smaller fragments.

Table 6.3 gives the results of examination of microhardness H_μ, the macroscopic limit of elasticity σ'', ultimate strength σ_B, yield limit $\sigma_{0.2}$ and relative elongation to fracture Δ of the titanium specimens in different conditions prior to and after ultrasound deformation.

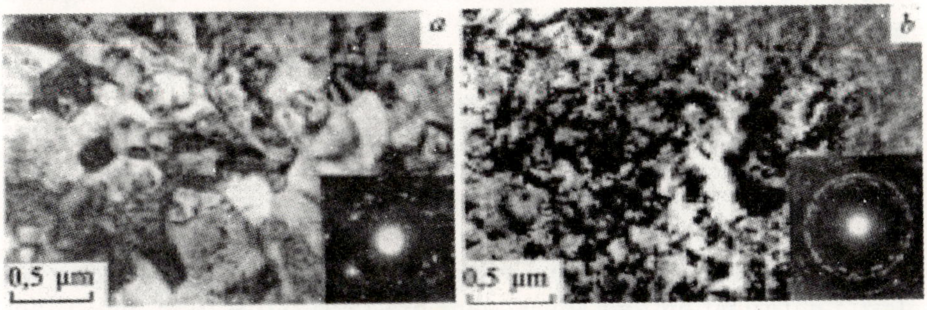

Fig. 6.7. The microstructure of the surface layer of nanostructured titanium prior to (*a*) and after (*b*) ultrasound deformation.

Table 6.3. Properties of coarse-grained and nanostructured titanium prior to and after ultrasound deformation

Material	Condition	H_μ	σ''	$\sigma_{0.2}$	σ_B	Δ, %
		MPa				
Coarse-grained	Initial	1800	180	370	470	21.9
	After USD	2750	240	480	530	13.4
Nanostructured	Initial	2300	260	650	710	18.9
	After USD	3170	270	650	710	13.2
Cold-worked	Initial	3090	220	1040	1090	5.0
	After USD	3515	230	1020	1060	5.8

Ultrasound deformation increases the microhardness of all investigated specimens. The maximum increase (by a factor of 1.5) is recorded in the initial recrystallised titanium VT1–0, and the weakest effect is recorded after severe deformation.

Ultrasound deformation results in different changes of the mechanical properties for coarse-crystalline and nanostructured titanium (Fig. 6.8). In coarse-grained titanium the values of σ_B and $\sigma_{0.2}$ increase as a result of ultrasound deformation, but the total elongation to fracture greatly decreases. In the initial and deformed conditions, nanostructured titanium did not show any significant changes in the yield limit and ultimate strength. The relative elongation to fracture in the case of nanostructured titanium decreased and in the deformed condition this parameter was almost constant. It should be noted that the form of the yield curve after ultrasound treatment also remained constant for all investigated specimens (Fig. 6.8).

The macroscopic yield limit σ'', characterising the variation of the law of build up of plastic strain in the region of micro-deformation in the case of recrystallised titanium VT1–0 after ultrasound deformation increase, and in the nanostructured condition, ultrasound treatment had almost no effect on the value of σ''.

In [198, 307] it was shown that the highest values of the endurance limit are shown by nanostructured titanium after cold plastic deformation and subsequent pre-recrystallisation annealing. Therefore, in [308] fatigue testing was carried out only on the specimens of nanostructured titanium with this structure. For this

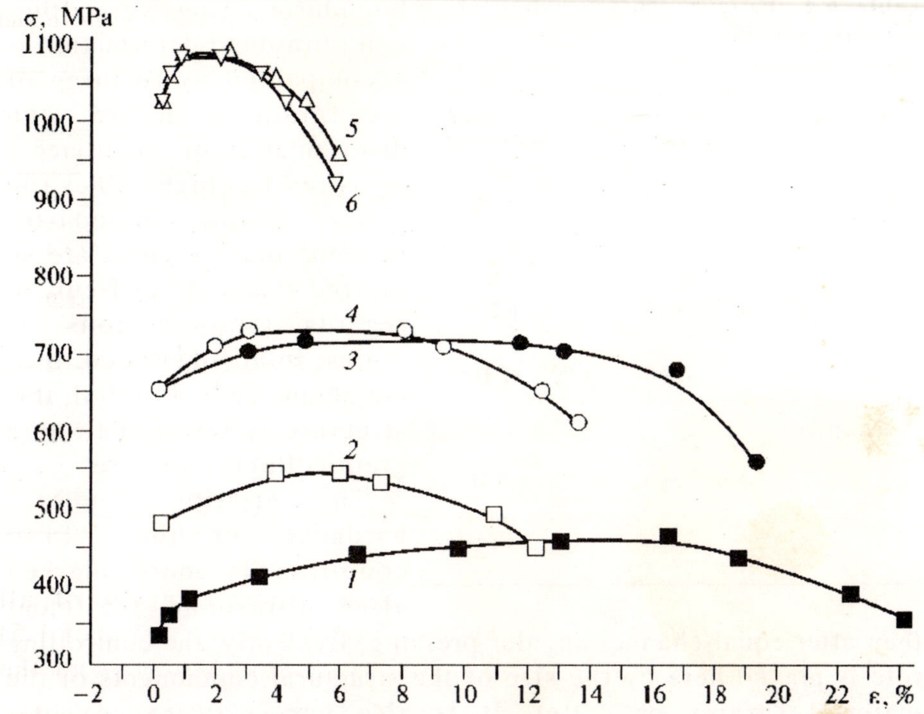

Fig. 6.8. Tensile curves of coarse-grained titanium VT1–0 (*1, 2*) and nanostructured titanium after equal-channel angular pressing (*3, 4*) and cold deformation (*5, 6*) prior to (*1, 3, 5*) and after (*2, 4, 6*) ultrasound deformation of the surface.

condition, the endurance limit for finite life at 106 cycles in the given loading method [198] is 520 MPa. To reduce the test time, the maximum cycle stress was varied in the range 520–625 MPa. The results of these tests are presented in Table 6.4 which shows that the longevity of the material is almost constant as a result of ultrasound deformation. On the basis of the data it may also be assumed that the fatigue limit for finite life did not show any significant changes.

The experimental data show that ultrasound deformation of the surface has the strongest effect on the properties of the initial recrystallised titanium VT1-0. The structural examination and the data presented in Table 6.3 make it possible to compare the efficiency of the effect of ultrasound surface treatment and severe plastic deformation by the method of equal-channel angular pressing on the properties of this material. The microstructure after this type of treatment differs. After equal-channel angular pressing, the material contains ultrafine grains with predominantly high-angle grain

Table 6.4. Fatigue characteristics of nanostructured titanium

Condition	Initial maximum cycle stress$_{max}$, MPa	Number of cycles of fracture N
Initial	520	$>10^6$
	550	$9 \cdot 10^4$
	625	$6 \cdot 10^4$
After ultrasound deformation	560	$12 \cdot 10^4$
	580	$6.5 \cdot 10^4$
	590	$3 \cdot 10^4$

boundaries. However, although ultrasound deformation is accompanied by refining of the grain structure and disorientation of the adjacent areas reaches high values (up to 40°), the boundaries between these regions are so blurred that it is difficult to regards these regions as grains; in all likelihood, they are unique cells. In [308], this structure is referred to as a grain-cellular structure.

The variation of microhardness for the VT1–0 condition is more marked after ultrasound treatment than after equal-channel angular pressing. Evidently, the controlling role is played here by the size of the structural components of the material (grains or cells). It is also necessary to consider differences in the nature of distribution, and the level of mean residual stresses.

When analysing the changes in the mechanical properties, it is necessary to take into account of that ultrasound deformation does not result in changes in the entire volume of the material and only in changes in the subsurface layers. Taking into account the fact that the thickness of the layer modified by ultrasound deformation is approximately 100 μm with the thickness of the specimens for mechanical testing of 400 μm, then even for the same efficiency of the effect of equal-channel angular pressing and ultrasound treatment, the variation of $\sigma_{0.2}$ and σ_B in the latter case will be half the value in the previous case. The data in Table 6.3 confirmed this assumption, although in slightly different proportions. At the same time, the relative elongation to fracture after ultrasound deformation is considerably greater than after equal-channel angular pressing.

Thus, ultrasound the formation of the surface of titanium specimens in the recrystallised coarse-grained condition leads to the formation of the microstructure in the surface layer which differs from the microstructure after the effect of equal-channel angular pressing, but as regards the efficiency of the effect on the investigated mechanical characteristics, it is fully comparable with the formation of the nanostructured condition in the material.

However, if ultrasound deformation is carried out on the titanium specimens with the nanostructure already produced by the method of equal-channel angular pressing, the properties of the material change in a different direction. The volume plastic deformation and ultrasound deformation of the surface of resulted in approximately the same change of microhardness and the macroscopic limit of elasticity. At the same time, ultrasound treatment has almost no effect on the yield limit and ultimate strength, whereas volume deformation greatly increases the is characteristic. In this case, volume deformation also has a stronger effect on the relative elongation to fracture, i.e. ultrasound deformation of nanostructured titanium is accompanied by the most intensive changes of the properties of the material, which depend on the surface condition.

The data published in [197, 198, 295, 307] indicate that the city properties of titanium in both coarse-grained and nanostructured conditions are in certain correlation with the strength characteristics of the material: as the ultimate strength increases, the endurance limit also increases. Consequently, this may indicate that the longevity of nanostructured titanium under cyclic loading and its endurance limit do not change greatly as a result of ultrasound deformation, because the characteristics of the yielding curve remained almost completely unchanged. The experiments carried out in [308] confirmed this assumption (Table 6.4).

6.3. GRAIN BOUNDARY INTERNAL FRICTION

It is generally accepted that the grain boundary peak of low frequency internal friction is determined by the true (natural) grain boundary sliding [288, 316–318]. In the most widely used model, the shearing of one grain on anlother in the plane of the boundary takes place by sliding of grain boundary dislocations. The conservative motion of the dislocations along the grain boundary under the effect of stress should be accompanied by the non-conservative motion of the dislocations which requires transfer of matter. This takes place by grain boundary and volume diffusion and, consequently, the activation energy of the true grain boundary sliding is higher than the activation energy of grain boundary diffusion but lower than that for volume diffusion.

Self-diffusion and diffusion of the substitutional atoms along the grain boundaries are considerably faster than in the body of the grain, because of the special features of the structure of the layer with a thickness of (2–4)a along the boundary with the flow of matter taking place along the boundary (here a is the lattice spacing) [288]. The activation energy and coefficients of grain boundary diffusion of the substitutional atoms and self diffusion depend on the orientation and type of grain boundaries, the structure (defectiveness) and the degree of nonequilibrium.

For nanostructured metals, produced by equal-channel angular pressing, the characteristic features are highly non-equilibrium grain boundaries with a specific defective structure (they contain steps, facets, dislocations and other defects), a large free volume, and high elastic deformation and long-range stresses in the vicinity of the boundary [150]. Pre-recrystallisation annealing of the metals with this type of structure results in the recovery processes at the grain boundaries which increase the degree of equilibrium of the boundaries, i.e. the activation energy of grain-boundary self diffusion should increase and the coefficient of grain boundary self-diffusion should decrease. Therefore, on the basis of the displacement of the grain boundary peak of internal friction in the range of high temperatures it is possible to evaluate the development of recovery processes at the grain boundaries.

The authors of [319] investigated the effect of pre-recrystallisation annealing on the grain-boundary internal friction of nanostructured titanium, whose structural-phase condition is shown in section 4.1. The coarse-grained structure was produced by annealing of this type of titanium at a temperature of 1023 K for 1 h. Annealing was carried out in a vacuum of 10^{-4} mm Hg. Internal friction was measured at an oscillation frequency of approximately 1 Hz, with the specimen heated at a rate of 400 deg/h.

The investigated temperature dependence of internal friction [319] showed the following special features of the scattering of the energy of elastic oscillations of the game boundaries of nanostructured titanium in comparison with coarse-grained titanium. Firstly, the extent of scattering of the energy of elastic oscillations of the grain boundaries increases. Secondly, the temperature range of manifestation of grain boundary internal friction is displaced to lower temperatures by approximately 200 K (Fig. 6.9, *1, 2*). Consequently, it may be assumed that in transition from the coarse-grained to nanostructured condition the activation energy of true grain boundary sliding decreases.

According to [316–318], the rising section of the grain boundary

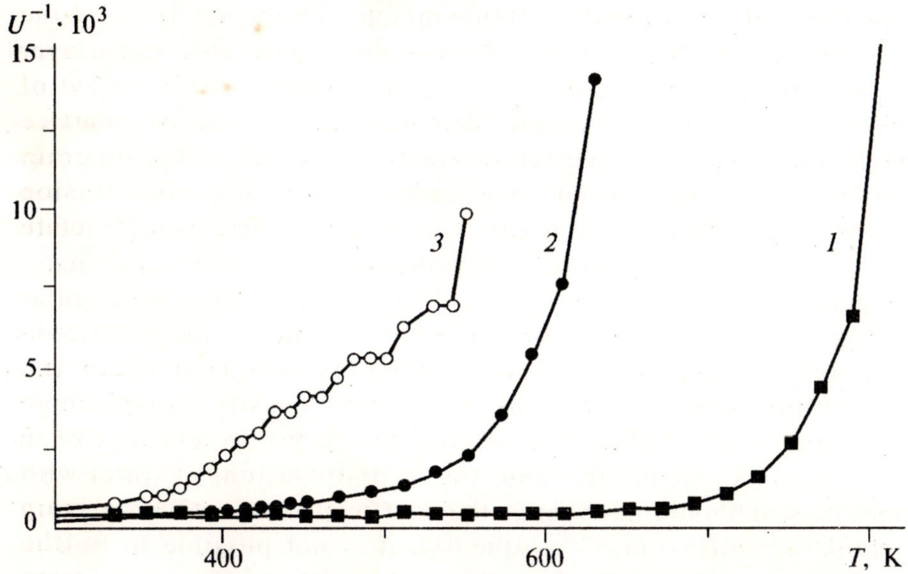

Fig. 6.9. Temperature dependence of internal friction for recrystallised titanium (*1*), nanostructured titanium after equal-channel angular pressing (*2*) and cold deformation (*3*).

peak of internal friction should be described by the equation

$$U^{-1} = U_0^{-1} \exp(-Q/RT),$$

where U_0^{-1} is a constant; R is the universal gas constant; T is temperature; Q is the activation energy of true grain boundary sliding. Analysis of the temperature dependence of internal friction shows that the rising part of grain boundary internal friction for both coarse-grained and nanostructured titanium is indeed described by the above equation. Consequently, on the basis of the angle of inclination of the ln U^{-1}–$1/T$ straight lines it is possible to determine the activation energy of grain boundary sliding Q. The results show that in transition from a coarse-grained nanostructured condition the energy decreases from 170 to 70 kJ/mole. The activation energy of the grain boundary sliding in the case of a coarse-grained titanium is higher than the activation energy of grain boundary diffusion (97 kJ/mole) but is lower than the activation energy of volume diffusion (242 kJ/mole) [207]. This is in agreement with the data for a large number of polycrystalline metals [318].

To explain the stability of the structure of the grain boundaries of nanostructured titanium, the same specimen was subjected to gradual annealing at different temperatures with a step of 50 K directly in the equipment for measuring internal friction. After each annealing, the temperature dependence of internal friction was investigated, and the maximum temperature measurement of internal friction did not exceed the temperature of subsequent annealing. The experimental results show that with an increase of annealing temperature, when this temperature was 573 K and higher, the grain boundary equal to internal friction was displaced to the region of higher temperatures (Fig. 6.10). This indicates the increase of the activation energy of grain boundary sliding with increasing annealing temperature.

Equal-channel angular pressing for producing nanostructured titanium was applied at elevated temperatures (not lower than 573 K) and, consequently, the dislocation density should not be excessively high. Although annealing does result in a change of the dislocation substructure and the redistribution of interstitial impurities, it should not change the coefficient and activation energy of bulk self-diffusion. Consequently, it is not possible to link the displacement of the grain boundary internal friction peak to higher temperatures with the increase of annealing temperature as a result of variation of the bulk diffusion coefficient. It remains to be

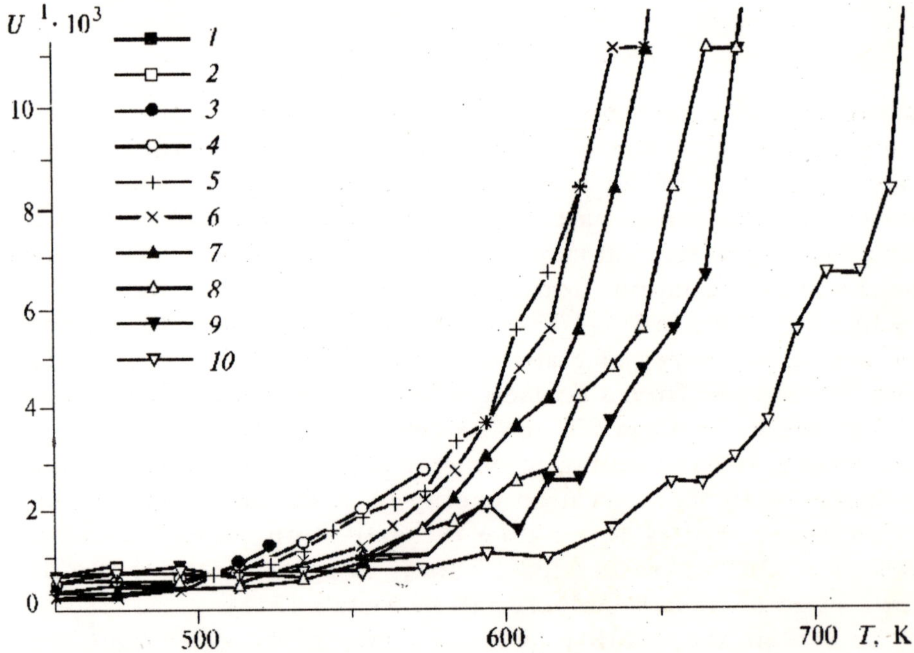

Fig. 6.10. Temperature dependence of the internal friction of nanostructured titanium after annealing for 30 minutes at different temperatures. *1*) initial; *2*) at T = 423 K, *3*) 473, *4*) 523, *5*) 573, *6*) 623, *7*) 673, *8*) 723, *9*) 773, *10*) 823 K.

assumed that this displacement is caused by a decrease of the coefficient of grain boundary self-diffusion. These data indicate a decrease of the degree of nonequilibrium of the grain boundaries in annealing.

Cold plastic deformation of nanostructured titanium (deformation by rolling of 40% at room temperature) resulted in the displacement of the grain boundary peak at lower temperatures by approximately 70 K (Fig. 6.9, *3*). Annealing of the specimens by the same procedure as in the case of initial and nanostructured titanium, results in the displacement of the grain boundary internal friction peak to higher temperatures (Fig. 6.11). Prior to the start of recrystallisation of plastically deformed specimens, the rising part of the grain boundary internal friction peak was observed at lower at temperatures, and in the presence of recrystallisation it was found at higher temperatures in comparison with the initial specimens.

These results indicate that with increasing annealing temperature

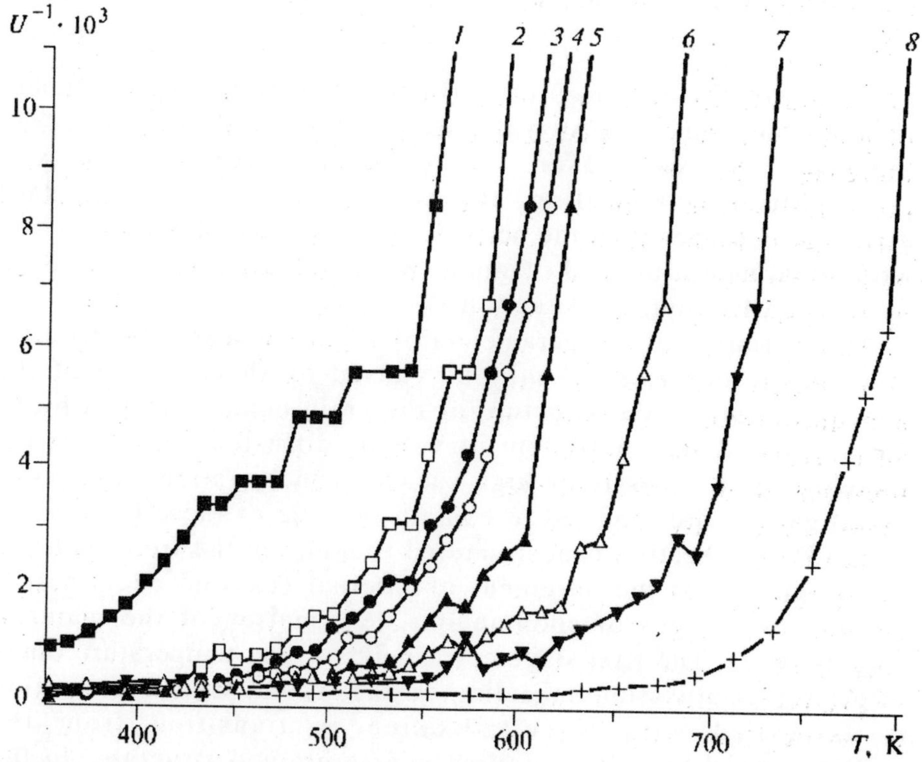

Fig. 6.11. Temperature dependence of the internal friction of nanostructured titanium, deformed at 293 K by rolling with a deformation of 40%, after annealing at different temperatures. *1*) initial; 2) at T = 573 K, *3*) 623, *4*) 673, *5*) 723, *6*) 773, *7*) 823, *8*) 873 K.

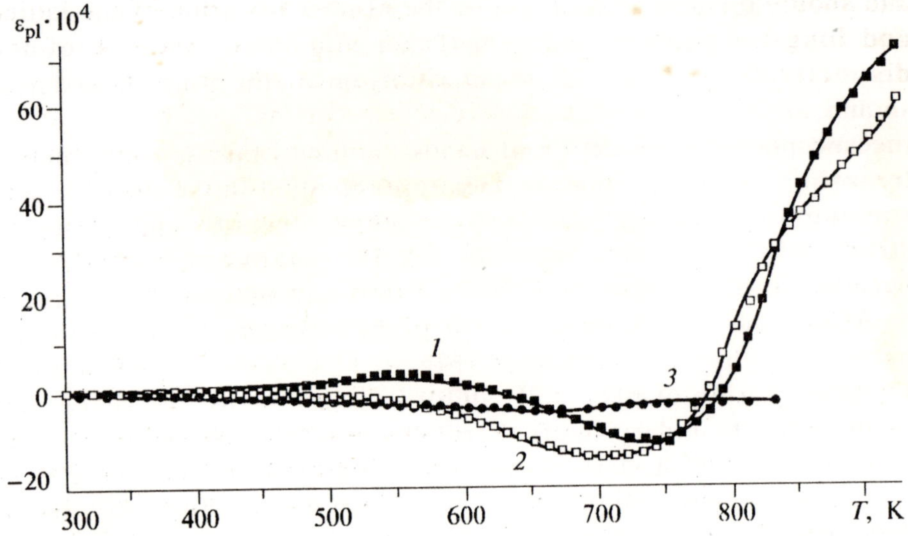

Fig. 6.12. Spontaneous deformation of nanostructured titanium in heating (*1*) cold-deformed nanostructured titanium in cooling (*2*), and titanium recrystallised in heating (*3*).

of the plastically deformed nanostructured titanium, the coefficient of grain boundary diffusion decreases, and the activation energy increases. Secondly, for the examined structure in pre-recrystallisation annealing, the coefficient of grain boundary diffusion is higher than the activation energy lower in comparison with initial nanostructured titanium. When annealing is accompanied by recrystallisation, the situation is reversed.

This variation of the parameters of grain boundary sliding after cold plastic deformation may be caused by the increase of the nonequilibrium of the structure of the grain boundaries and by an an increase of the coefficient of volume diffusion because of the presence of an excessively high vacancy concentration. Additional investigations are required to explain the role of these factors.

In [319] in heating nanostructured titanium in the free condition in equipment for measurement of internal friction, examination showed two stages of spontaneous deformation of the material (Fig. 6.12, *1*). The first stage corresponds to the temperature range of pre-recrystallisation annealing, in the second stage deformation is associated with recrystallisation in transition from the nanostructured to polycrystalline coarse-grained structure. In the first stage of spontaneous deformation, as indicated by the results of measurements of grain boundary internal friction, the degree of nonequilibrium of the structure of the grain boundaries decreases and should result in a decrease of the elastic distortion of the lattice and long-range stresses at the grain boundary. Because of the distinctive texture [150], these changes at the grain boundaries should result in spontaneous deformation of the specimen which in the second stage is evidently associated with the changes of the texture in recrystallisation. The spontaneous deformation of the specimen in the first and second stage is caused by different structural factors, as indicated by the change of the sign of deformation in transition from the first to second stage.

After cold plastic deformation of nanostructured titanium, the nature of the temperature dependence of spontaneous deformation of this material does not change (Fig. 6.12, *2*). Only the temperature of the start of the second stage decreases and this is in agreement with a decrease of the temperature of the start of recrystallisation. After recrystallisation annealing (1023 K) of nanostructured titanium, spontaneous deformation does not take place when the temperature is changed (Fig. 6.12, *3*).

CHAPTER 7

SUPERPLASTICITY OF NANOSTRUCTURED MATERIALS

According to theoretical estimates, from the viewpoint of mechanical behaviour, the formation of nanostructures in different metals and alloys may result in the formation of the high-strength state in accordance with the Hall–Petch equation, and also in low temperature and/or high-rate superplasticity. The realisation of these possibilities is of direct importance for the development of new high-strength and wear-resistant materials, promising superplastic alloys and metals with high fatigue strength. Consequently, there is a considerable interest in research into the strength and plasticity of the materials in order to produce large volume specimens with the nanostructure for subsequent mechanical testing.

Many of the previously mentioned problems of the formation of specimens can be overcome by the development of nanostructures in coarse crystalline materials as a result of the application of methods of severe plastic deformation, based on the deformation of specimens with higher degrees of reduction at relatively low temperatures (usually lower than 0.3–0.4 T_m) in the conditions of high applied pressures [150, 320]. The materials produced by severe plastic deformation often show superplasticity at relatively low temperatures and may demonstrate high-rate superplasticity (more information on this will be provided in chapter 7.2).

In the present chapter, the results are presented of the examination of the effect of special features of the structure on the superplastic properties. Problems of the mechanisms of the formation of other materials and methods of producing unique properties are discussed.

7.1. PHENOMENOLOGY OF SUPERPLASTICITY

The conditions of transition to the superplastic condition are determined by three main factors: the presence of a fine-grain structure (d <10–15 μm), deformation temperature usually higher than 0.5 T_m and the rate range of deformation (usually 10^{-4}–10^{-3} s^{-1}). In almost all metals, the superplasticity effect has common features and, consequently, it can be regarded as a special type of plastic deformation, characterised not only by the specific phenomenology but also by the strong effect of the structure on the mechanical properties of metallic materials.

The largest difference between conventional deformation and superplasticity is that in the former the flow stress depends strongly on the degree of strain and relatively only slightly on the strain rate. In the superplastic condition, the situation is reversed. The flow stress depends only slightly on the strain but strongly on the strain rate. Consequently, structural superplasticity may be determined as the capacity of metals and alloys for high plastic strains without failure in the conditions of high rate sensitivity of the flow stress. In the nanostructured materials, characterised by the formation of the ultra fine-grained microstructure (d <1 μm) the appearance of high-rate and/or low temperature superplasticity is highly likely.

Phenomenology of the phenomenon and its relationship with the initial microstructure

The phenomenology of the behaviour materials in the superplastic conditions has already been described in detail [138, 139, 271, 282, 321–323]. According to [139, 285], we present the most important results of examination of the phenomenology of superplastic flow essential for understanding the mechanism of deformation and the properties of typical superplastic alloys.

One of the most characteristic features of superplastic deformation is the high deformation resistance of the specimen, i.e. the resistance to the formation and growth of a neck during tensile loading. In accordance with the assumptions, published by Hart, this stability may be understood on the basis of the dependence of the flow stress on the strain and the strain rate which has the following form for the isothermal conditions:

$$\sigma = A \cdot \varepsilon^q \cdot \dot{\varepsilon}^m, \quad (7.1)$$

here A is the empirical constant; q, m are the parameters of the material determining the dependence of hardening on the strain and strain rate.

For the materials with a weak dependence of the flow stress on the strain rate, i.e. at m close to zero, the strain hardening parameter may be described by the following equation:

$$q = d\lg\sigma / d\lg\varepsilon. \tag{7.2}$$

For materials showing slightly hardening in deformation, equation (7.1) has the following form:

$$\sigma = A \cdot \dot{\varepsilon}^m, \tag{7.3}$$

and the rate hardening parameter is determined from the equation:

$$m = d\lg\sigma / d\lg\dot{\varepsilon}. \tag{7.4}$$

The stability of plastic flow is estimated using these parameters, examining the conditions of formation and growth of a neck in the specimen. In the area of localisation of deformation, the flow speed and the degree of deformation increase in comparison with the remaining part of the specimen. Correspondingly, this part of the specimen is hardened and the hardening is determined by the parameters q and m. Thus, the high stability of the plastic flow may be realised as a result of deformation cold working or rate hardening of the material, i.e. high values of the coefficients q and m.

It is well-known that the conventional plastic flow is accompanied by a continuous hardening during tensile loading and this is followed by localisation of deformation, formation of a neck and fracture. In addition to this, as shown by the analysis of the true tensile curves, in the superplastic condition, deformation already at low degrees transfers to the stages in which the plastic flow develops under the effect of low and constant stresses almost without hardening to hundreds and thousands of percent of elongation, and sometimes even softening may be detected. Localisation of deformation in the form of necking is either arrested or its rate is low throughout the entire process of tensile loading.

The rate dependence of the mechanical properties of superplastic alloys also greatly differs from conventional deformation (Fig. 7.1). Superplastic deformation is characterised by the strong dependence of the flow stress on the strain rate which is sigmoidal on the logarithmic scale (Fig. 7.1a). Because of the sigmoidal form

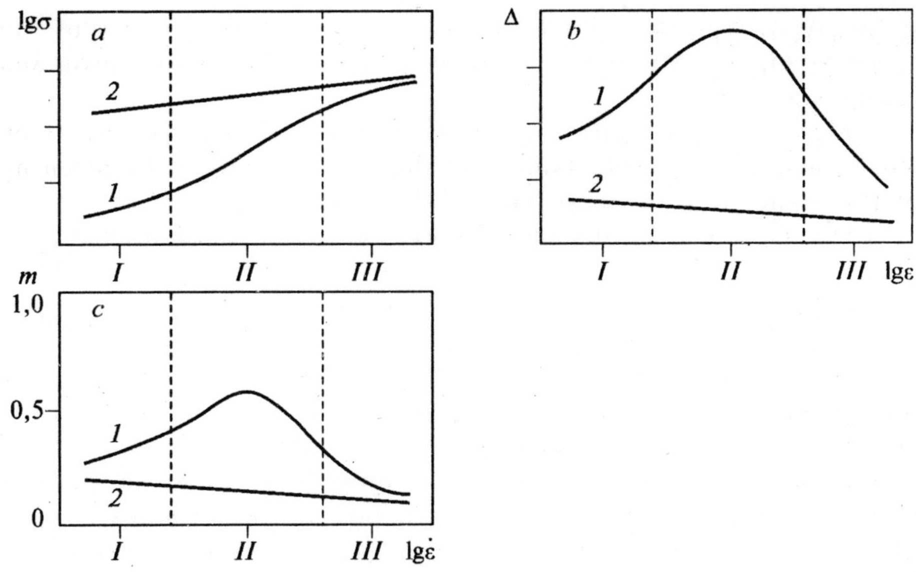

Fig. 7.1. Typical dependences of flow stress δ (*a*), relative elongation Δ (*b*) and coefficient *m* (*c*) on the strain rate of the alloys in the superplastic (*1*) and normal (*2*) conditions.

of the lg (σ)–lg $\dot{\varepsilon}$ curve, three sections can be found. At low strain rates, the dependence of lg σ on lg $\dot{\varepsilon}$ is relatively unpronounced (Fig. 7.1*a*, region I) with low values of *m* and Δ (Fig. 7.1*b*, *c*). With increasing $\dot{\varepsilon}$, the dependence of the flow stress on the strain rate becomes 'sharper', the values of *m* and Δ increase and the transition takes place to the range II where the superplastic effect reaches the maximum value. The range of the strain rate, corresponding to the region II, slightly differs for different alloys but is usually in the range 10^{-4}–10^{-2} s^{-1}. A further increase $\dot{\varepsilon}$ of reduces the values of *m* and Δ. In region III at high values of $\dot{\varepsilon}$, the relative elongation, flow stress σ and the parameter *m* approach the values characteristic of conventional plastic materials. This form of the curve lgσ–lg$\dot{\varepsilon}$, often referred to as the superplastic curve, showing clearly the three regions, is typical of different materials in the superplastic condition.

It should be mentioned that the mechanical model of the viscoplastic flow, developed in the studies by Hart, makes it possible to explain only qualitatively the behaviour of superplastic materials and the attempts to determine the quantitative relationship between the value of coefficient m and plasticity have not provided unambiguous results [138, 322, 323]. Nevertheless, as already noted, almost all alloys in the superplastic condition showed the qualitative relationship between the value Δ and coefficient *m* which is the most important characteristic of the superplastic materials.

Thus, superplastic deformation is characterised by the following features:

– high values of the relative elongation of the specimens in tensile loading and high stability of the plastic flow process because of the higher rate of sensitivity of the flow stress;

– low flow stress, absence of any significant hardening during plastic flow;

– dependence of the flow stress, elongation and parameter m on the strain rate.

The parameters of superplastic flow σ, Δ, m and the position of the optimum rate range are determined by the structure of the alloy and deformation temperature. Analysis of the effect of different factors of superplastic deformation is presented below. The factors that are independent of the microstructural characteristics will be discussed briefly and special attention will be given to the effect of the mean grain size on the superplastic flow.

Temperature

Superplastic deformation is characterised by the strong dependence of the mechanical properties on deformation temperature. The increase of temperature reduces the flow stress and results in the shift of the optimum rate range of superplasticity (Fig. 7.1, region II) to higher values of $\dot{\varepsilon}$. However, this effect of temperature is detected only in the range in which the stability of the ultrafine-grained structure is retained. If the increase of temperature is accompanied by a rapid grain growth or complete loss of stability of the structure as a result of transition of the metal to the single-phase state, the strength of the superplastic effect rapidly decreases or the effect completely disappears. The dependence of the rate of superplastic deformation on temperature may be described by the equation:

$$\dot{\varepsilon} = M\exp(-Q/RT), \tag{7.5}$$

here Q is the activation energy; R is the gas constant.

The identical dependence may be used for describing the effect of temperature on the flow stress:

$$\sigma = H'\exp(Q/RT). \tag{7.6}$$

The activation energy reflects the nature of thermally activated processes, controlling the strain rate, and is an important parameter of superplastic flow. The value of Q is usually determined by tests at a constant stresses or constant $\dot{\varepsilon}$ in a specific deformation temperature range.

Experimental calculations of the activation energy are carried out on the basis of the tangent of the angle of inclination of the dependences $\lg\dot{\varepsilon} - 1/T$ or $\lg \sigma - 1/T$, respectively, for the individual test conditions. However, the pre-exponential multipliers in equations (7.5) and (7.6) usually depend on temperature and, therefore, in principle, the method makes it possible to measure the so-called apparent activation energy (not true activation energy) which may slightly differ from the true value of Q. Regardless of some conventionality, the measurements of the activation energy are of great importance for the analysis of the nature of superplastic flow and this was estimated in [243, 323] and other studies. In region II, the value of Q is usually close to the activation energy of grain boundary diffusion but increases with an increase of the activation energy and with a decrease of the strain rate, and in the region II it is often close to the activation energy of volume diffusion.

Grain size

In the group of the structural factors influencing the superplasticity effect, it is important to mention in particular the grain size. The stable ultrafine-grained structure with d <10–15 μm is an essential condition for the formation of structural superplasticity.

On the example of alloys close to single-phase alloys (because in these alloys it is possible to produce equiaxed grains from the units to hundreds of micrometres in size), we can study the transition from conventional deformation to superplastic flow with the refining of the microstructure. For example, for the Zn–0.4 % Al alloy it has been shown [243] that at d > 10 μm, the dependence of $\sigma_{0.2}$ on d (Fig. 7.2) is governed by the Hall–Petch relationship, in accordance with which $\sigma \sim d^{-1/2}$. However, at d<10 μm the dependence of the yield limit on the grain size changes to the reversed dependence and the value of $\delta_{0.2}$ decreases with decreasing d.

Refining of the structure is also accompanied by changes in the plasticity of the Zn–0.4% Al alloys and also the rate dependence of the properties of the alloy. Especially large changes of Δ were recorded with the variation of the grain size from 10 to 1 μm, i.e. relative elongation increases from 25 to 400% (Fig. 7.2) and at the same time, the dependence of the properties on the strain rate, characteristic of superplasticity, becomes more marked.

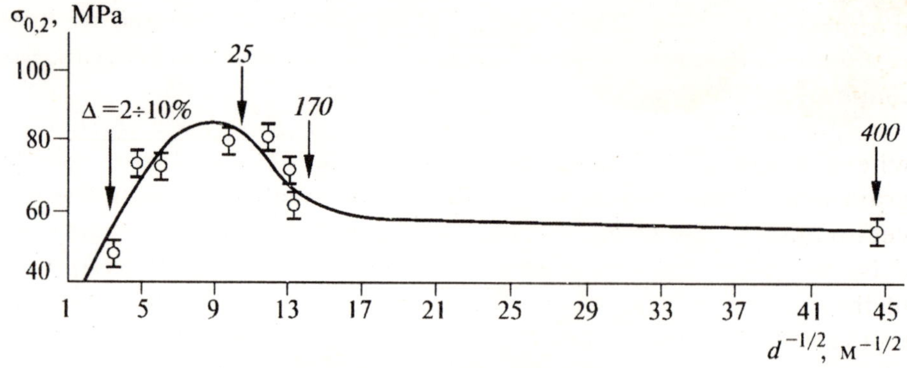

Fig. 7.2. Dependence of the yield limit ($\sigma_{0.2}$) of the specimens of Zn–0.4% Al alloys on the grain size (d) at 293 K and $\dot{\varepsilon} = 10^{-3}$ s^{-1}.

The authors of [138, 243, etc.] investigated the quantitative dependence of the stress of superplastic flow on grain size. It has been shown that in superplastic deformation, the dependence has the form:

$$\sigma \sim d^{-p}, \tag{7.7}$$

where p = 0.7–2, and in most cases p = 1.

It has been established that at constant σ the effect of the grain size on the strain rate in the second region may be expressed by the relationship:

$$\dot{\varepsilon}_{II} \sim d^{-j}, \tag{7.8}$$

where j = 2–3.

The change of the grain size has also a strong effect on the dependence of the flow stress on the strain rate: a decrease of d displaces the optimum range of $\dot{\varepsilon}_{II}$ in which the superplastic effect becomes evident, to higher strain rates, usually by several orders of magnitude. Correspondingly, the variation of the grain size displaces the maxima of the values of the $m-\dot{\varepsilon}$ and $\Delta-\dot{\varepsilon}$ relationships to higher values of $\dot{\varepsilon}$, and the maximum value of increases.

Thus, the stable fine-grained structure is the essential condition for producing the superplastic effects. At present, this has been established for the large majority of superplastic alloys. It is regarded as generally recognised that the superplasticity may be described by a generalised equation linking the flow stress and the strain rate [322]:

$$\dot{\varepsilon} = A\frac{DGb}{kT}\left(\frac{b}{d}\right)^p\left(\frac{\sigma}{G}\right)^n, \qquad (7.9)$$

where D is the appropriate diffusion coefficient (volume or grain boundary diffusion); G is the shear modulus; b is the Burgers vector; k is the Boltzmann constant; T is deformation temperature; d is the mean grain size; σ is the flow stress; p is the exponent of the dependence on the grain size; $n = 1/m$ is a quantity reciprocal to the rate sensitivity coefficient. The results of a large number of experiment show that in metals, intermetallics and ceramics with a nanostructure, the exponent p is close to 2 if the main accommodation mechanism is volume diffusion, and is equal to 3 in the case of grain boundary diffusion [282, 321, 322].

In many actual cases, the application of the equation of state is of limited importance because the superplastic behaviour may also be affected by other factors, such as the type of grain boundary, crystallographic texture, etc., which are not taken into account directly by the equation of state. Several such examples will be examined for superplastic materials.

In a number of cases, the features of superplastic flow were found in coarse-grained materials [324–328]. The reasons for this behaviour of the alloys differ. In particular, it has been established that the process of deformation of coarse-grained titanium alloys is accompanied by the formation of a substructure inside the grains, and the behaviour of the substructure is probably similar to the fine-grained structure of superplastic materials [10]. At the same time, examination of the alloys of the Al–Ge systems shows that the superplasticity effect is found at d = 100–200 μm [327–329]. In this case, the formation of the substructure during deformation was not detected. At the same time, a correlation was found between the initial porosity of the alloys, formed as a result of phase transformation in heating to test temperature, and the relative elongation in the conditions of superplastic flow [328]. Maximum plasticity was recorded for the Al–0.4% Ge alloy in which the initial porosity also reached the highest value (approximately 0.8%) (volume fraction). It was shown in [329] that the superplastic effect in the Al–Ge alloys is caused by the fact that porosity supports the development of the combination of mechanisms characteristic of conventional superplastic alloys, and since the porosity is maintained constant, it does not cause fracture of the material.

Differences in the grain size

The effect of structural characteristics of the superplastic effect is usually analysed using the mean grain size, although in individual cases the size of the crystals in the tested alloy may differ by an order of magnitude or more. In particular, this relates to industrial alloys in which the formation of the fine grains in the process of severe preliminary plastic deformation is sometimes accompanied by the formation of a zone of large grains, in addition to the fine-grained structure. The presence of differences in the grain size naturally changes the properties of the alloys in the conditions of superplastic flow because the formation of larger grains in the structure should reduce the intensity of the superplastic effect.

These considerations can be illustrated efficiently by the example [243, 282] of industrial aluminium alloy 01420 (Al–5 % Mg–1.8 % Li–0.12 % Zr). The alloy was produced in three conditions; with the equiaxed fine-grained microstructure d = 6 μm, with a coarse-grained microstructure d = 75 μm, and with a mixed structure in which the large elongated grains were accompanied by equiaxed fine grains with d = 6 μm. The area of the fractions of the fine and large grains in this condition was 1:1.

Table 7.1 gives the dependence of σ, Δ and m on the type of microstructure in deformation of 1420 alloy (T = 723 K, $\dot{\varepsilon} = 10^{-3}$ s^{-1}). These data show that in the presence of a mixed structure the alloy also shows features of the superplastic condition. However, this is accompanied by a small increase of the flow stress at a decrease of a relative elongation. At the same time, the rate range of superplastic deformation is displaced to lower values of $\dot{\varepsilon}$.

Phase composition

The grain size is an important but not the only characteristic of the

Table 7.1. Dependence of the parameters of superplastic flow in 1420 alloy on the mean grain size and the dispersion of the microstructure

Superplasticity parameter	μm		
	6	75	6 and 75 (1:1)
Elongation, %	720	210	390
Flow stress, MPa	5	20	13
Rate sensitivity coefficient	0.55	0.22	0.40

structure which determine the properties of the superplastic alloys. The chemical and phase compositions have a strong effect on the structure and properties of the alloys in the superplastic flow conditions. The relationship of chemical composition with the superplastic effect was examined in greater detail in analysis of the effect of alloying on the properties of specific industrial alloys. It is assumed that the chemical composition has an indirect effect on superplastic deformation through the microstructure, i.e. by the formation of conditions for producing a stable ultrafine grain microstructure. The effect of phase composition on the superplastic effect is usually studied primarily from the viewpoint of the stability of the microstructure [139, 282, 321]. Actually, it is the most evident aspect of the effect of phase composition on superplastic deformation. Superplasticity was also detected and examined in alloys with approximately the same ratio of eutectic phases or eutectoid composition. In these materials, it is very easy to produce ultrafine grains and ensure that these grains are stable. At high deformation temperatures, grain growth is rapidly activated and this prevents the manifestation of the superplasticity effect in the pure metals and alloys [330]. It is obvious that with an increase of the amount of the second phase the stability of the structure of the alloys increases and, therefore, the multiphase alloys can be transferred more easily to the superplastic condition in comparison with the single-phase alloys, and the formation of the superplastic state in the special purity metals on the principle of structural superplasticity is almost impossible to achieve. However, the effect of phase composition on superplastic deformation is not exhausted only by the stabilisation of the microstructure and grain size of the alloys and the deformation characteristic of the phases are also very important. Thus, the role of the phase composition of the alloys is not only the formation of the optimum microstructure with a high stability of the grain size at $T > 0.4\ T_m$. When developing identical microstructures, it is also necessary to take into account the deformation characteristics of every phase, and also the phase interaction in the deformation process.

Texture

The effect of texture on the properties of metals and alloys is traditionally linked with the appearance of preferential orientation of the grains in a polycrystal. The mechanical properties of single crystals depend on the orientation of the crystal lattice in relation to the acting force and, consequently, in the presence of preferential orientation of the grains in polycrystals, the properties are anisotropic.

This effect for the conditions of superplastic deformation was examined for the first time on the Zn–0.4 % Al alloy [331]. It has been established that σ, Δ and coefficient *m* greatly change in relation to the direction of cutting out the specimen in relation to the rolling direction. According to the authors of [331] this fact is associated with the presence of a crystallographic texture in the Zn–0.4% Al alloy. Subsequently, the anisotropy of the properties in the conditions of superplastic deformation was detected in many alloys: Zn–Al [332–334], aluminium bronze [335], brass [336], Sn-Bi [337], Ti–6% Al–4% V [338]. However, the results are not always interpreted as a consequence of the presence of preferential orientation of the grains. For the metals and alloys, especially industrial materials, showing the superplasticity effect, the specific features of the preparation of the structure and the presence in the structure of impurities and inclusions result in a specific degree of heterogeneity of the structure: the elongation of the grains and the direction of distribution of inclusions and the phases which may influence the superplasticity effect. According to many authors, the anisotropy of the properties forms as a result of the directional distribution of inclusions and particles of the second phase in the structure.

It should be mentioned that anisotropy is often found only in the case of relative elongation and the flow stress is isotropic [333, 334, 338]. For example, in the Zn–22% Al alloy [333, 334], regardless of the presence of a texture, the anisotropy of the flow stress was not found and the dependence of the flow stress on the direction of cutting out the specimen was found only in alloys with the grains elongated in the rolling direction. The absence of the anisotropy of the flow stress complicates to a certain degree the interpretation of the effect of texture on the properties of alloys and some investigators, for example, the authors of [338] therefore assume that the preferential orientation of the grains does not change the superplasticity effect. However, the absence of the anisotropy of the flow stress may be a specific feature of the effect of texture on the process of superplastic deformation and does not indicate the absence of such an effect.

The texture influences the position of the optimum rate range of superplasticity. For the same microstructure, the presence of a texture in the material reduces the flow stress, increases relative elongation and shifts the optimum range of the strain rate to higher values of $\dot{\varepsilon}$. With decreasing test temperature these effects of the texture are not only retained but also intensified.

On the basis of the experimental results it may be concluded that the established changes of the characteristics of superplastic flow are evidently associated with the fact that the presence of the crystallographic texture changes the structure of the grain boundaries in the polycrystal [338]. The relationship of the texture with the structure of the grain boundaries does not lead to any doubts because, taking into account the distribution function of the orientations, indicated by the texture data, it is always possible to construct the function of the distribution of the boundaries on the basis of the characteristic dimensions, in particular the angle of disorientation between the grains. These geometrical parameters determine the structure of the grains boundaries. The grain boundary processes play a specific role in the manifestation of superplastic behaviour.

Condition of the grain boundaries

The structure of the ultrafine grain superplastic materials is characterised by a large increase of the total length of the grain boundaries (the size of the grains characterises in fact a quantity inversely proportional to the area of the grain boundaries per unit volume). In this case, the grain boundaries are not only the geometrical surfaces, separating the grains with different orientation, but also an important structural element of the polycrystalline material, determining its properties. Unfortunately, insufficient attention has been paid to this problem in the examination of structural superplasticity. This is due mainly to the shortage of data on the behaviour of the grain boundaries in plastic deformation processes. However, recent advances in understanding the structure and properties of the high-angle grain boundaries [206, 339, 340] have made it possible to find a new approach to the formulation and analysis of the experiments directed at explaining the role of the condition of the boundaries in the manifestation of superplastic deformation.

It is interesting to examine the results of investigations of the superplastic behaviour of MA8 magnesium alloy with twins in the structure. This alloy, annealed after cold rolling, is superplastic at 673 K and $d \approx 0$ μm [282]. The twins were introduced into the structure of the alloy by cold deformation of the specimen. After this type of treatment, more than 75% of all grains contained strain twins and the number of the twins remained almost constant in heating to the temperature of superplastic deformation. The presence of twin interlayers can be regarded as additional refining of the structure, taking into account the boundaries of the twins, the size of the grains in the alloy is approximately 3 μm. It is well-known that a decrease of the grain size usually facilitates the realisation of the superplasticity effect and shifts the range in which these effects occurs to higher rates. However, according to the results of mechanical testing, the situation in the alloy containing twins is different: the flow stress increases and the maximum values of the plasticity parameters (Δ and *m*) are displaced to lower strain rates. Examination of the structure showed reasons for these changes in the mechanical properties. It has been established that no sliding takes place at the twin boundaries and, at the same time, twins result in the buildup of dislocations in the structure. The twin boundaries are of the high-angle type, but belong to special grain boundaries showing special properties: they are not sinks for dislocations. Thus, the experimental results directly indicate the important role of the structure of the boundaries in the manifestation of superplastic deformation.

In addition to the structure of the grain boundaries, the development of grain boundary processes is determined by the type and concentration of solutes and segregation. Therefore, it may be expected that they also have a strong effect on the development of superplastic flow.

The resultant data indicate the importance of the condition of the grain boundaries in the manifestation of superplastic flow. They show that the development of the fine-grained structure is the essential but insufficient condition for the realisation of the superplastic effect. It is also important to take into account specific features of the structure and condition of the grain boundaries and the behaviour (sliding, migration) in the microstructure of the superplastic materials which should result in the development of processes determining superplastic deformation.

7.2. THE SUPERPLASTIC BEHAVIOUR OF NANOSTRUCTURED MATERIALS

According to the equation of state for the superplastic flow, a decrease of the grain size should result in a large increase of the superplastic properties and establishment of superplasticity at relatively low temperatures and/or high strain rates. Therefore, the development of the methods of severe plastic deformation for producing nanostructured materials (chapter 1) has opened new possibilities for examining superplasticity in metallic materials and has also made it possible to start new systematic experimental investigations in this area [150, 320, 341]. Investigations started in two directions: the first – on volume specimens with a homogeneous structure and the grain size smaller than 1 μm (the level of sub-microcrystals), produced by equal-channel angular pressing or multiple forging, the second one – on nanocrystalline specimens with small geometrical dimensions (<15–20 μm) which were produced using the method of severe plastic deformation by torsion.

Attention will now be given to a number of achievements in this area and to the comparative characteristics of the superplastic behaviour of several ultrafine-grained alloys produced by two methods: equal-channel angular pressing and severe plastic deformation by torsion. Special attention will also be given to a recent investigation of the unusual superplastic behaviour of nanocrystalline nickel produced by electrodeposition.

Superplasticity of nanocrystalline alloys

The superplastic behaviour of nanocrystals has been examined mainly for the alloys because the nanostructures are usually characterised by low stability at elevated temperatures and in many cases nanocrystalline pure metals are not stable even at room temperature. The nanostructures in the alloys and in the metallic compounds are more stable. The structures were produced by severe plastic deformation in torsion in boron-alloyed intermetallic compound Ni_3Al (Ni–3.5% Al–7.8% Cr–0.6% Zr–0.02% B) [212] and 1420 aluminium alloy (Al–5.5% Mg–2.2% Li–0.12% Zr) [212, 345]. This method (chapter 1) has advantages in producing small specimens in the form of discs (diameter 12 mm, 0.5 mm thick) with a nanostructure. Specimens for mechanical tensile testing with the gauge length of 1 mm were cut by electrospark cutting from discs subjected to severe plastic deformation in torsion. The tensile test was carried out in specially prepared testing equipment, with computer control of the strain rate [212, 342]. After severe plastic deformation by torsion, the Ni_3Al intermetallic compound showed considerable refining of the structure. Two microscopes were used for examination of the microstructure of Ni_3Al: Jeol 200CX and Jeol ARM–1000. The first microscope with the accelerating voltage

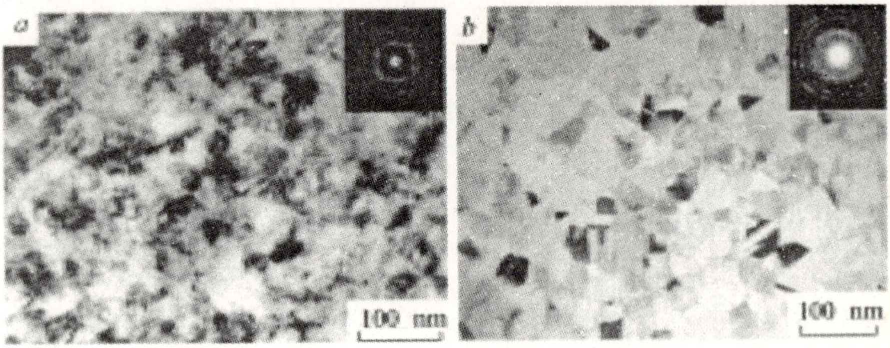

Fig. 7.3. Transmission electron micrographs of the structure of specimens of Ni_3Al subjected to torsion under pressure. *a*) after deformation, *b*) after annealing at T = 923 K, 30 min.

of 200 kV was used for examination of the grain structure, the second one with the accelerating voltage of 800 kV for the examination of the fine structure of the grain boundaries. Since the mean grain size of the specimens was smaller than 100 nm, the cross-section of the foil contained several grains. For examination by high-resolution microscopy, thinner sections were found with the grains mainly oriented in such a manner that the <100> direction in the lattice of the grains was normal to the plane of the foil. Figure 7.3*a* shows the typical microstructure of a twisted specimen together with the microdiffraction pattern (the diameter of the selector diaphragm 0.7 μm). Light field images, produced *in situ* in annealing, and the DSC (differential scanning calorimetry) curve are presented in Fig. 7.4. The nanocrystalline structure of the twisted specimens is stable up to a temperature of 773 K. No significant changes were found in the mean grain size. With an increase of the annealing temperature, the contours of the grain boundaries became sharper indicating that relaxation processes take place there. In heating to a temperature of 1023 K for 5 min, the microstructure of Ni_3Al remained nanocrystalline with the mean grain size of approximately 100 nm. At the same time, according to the results of differential scanning calorimetry, the temperature of 723 K is characterised by the formation of a distinctive exothermic peak, associated with generation of heat. In all likelihood, this peak does not correspond to the start of grain growth, it corresponds to recovery and ordering which were also detected previously in [174]. Figure 7.3b shows, at a high magnification, the microstructure of Ni_3Al after annealing at a temperature of 923 K for 30 min. It may be seen that the grains grew only slightly but the boundaries

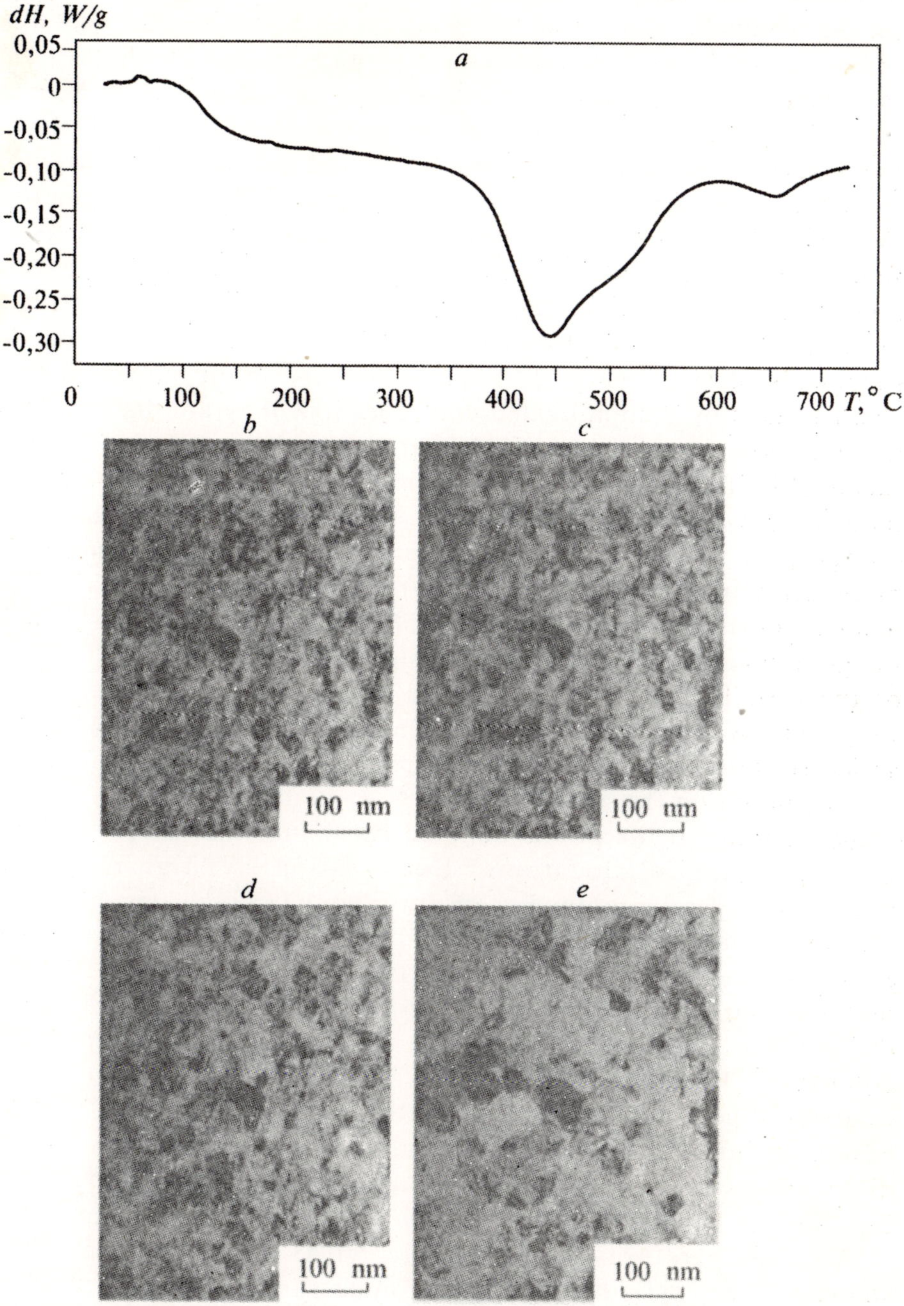

Fig. 7.4. DSC curves (*a*) and the microstructural microcrystalline Ni_3Al in *in situ* annealing (*b*-initial, *c*–3 min at T = 573 K, *d*–4 min and 773 K, *e*–5 min at 823 K).

became considerably sharper. The grain structure is more equiaxed but the residual distortion of the grains is still visible. More detailed examination of the microdiffraction pattern shows that in the specimens after annealing at a temperature of 923 and 1023 K there are extra-reflections, corresponding to the ordered Ni_3Al-phase. These reflections were not found on the microdiffraction pattern is from the twisted specimens and after low temperature annealing (573 K). Thus, it may be concluded that in heating the twisted specimens there is no considerable grain growth, the structure relaxes, remains nanocrystalline, but atomic reordering takes place in the lattice. The high-resolution electron microscopy of the specimens of Ni_3Al (Fig. 7.5) shows that the grain boundaries are characterised by a complicated configuration: they are distorted and the contours are not sharp. There are several regions in Fig. 7.5a (denoted by the letters A and B) with no sharp images of the planes indicating the local distortion of the lattice in the given areas. This coincides with the conclusions made in [343–345] where the nonequilibrium condition of the grain boundaries was found in nanocrystalline materials. Figure 7.5*b* shows another area in which the grain boundaries are almost completely perpendicular to the plane of the foil. It should be mentioned that the grain boundaries are very narrow, their thickness is smaller than 0.5 nm, which is approximately equal to two planar spacings. However, the boundary is not smooth on the atomic level, i.e., contains small steps. Lattice planes, ending at the points denoted by T, indicate the presence of grain boundary dislocations. Examination by high-resolution microscopy of a large number of grain boundaries in the specimens of Ni3Al, deformed in torsion under pressure and subsequently annealed at 923 K, indicates that after annealing partial recrystallisation took place in the grain boundaries but in all cases the grain boundaries were in the nonequilibrium, high-energy condition. Figure 7.5*c* shows a ternary joint in the Ni_3Al intermetallic compound after annealing at 1023 K (3 min). In this case, the grain boundaries are small but the density of the dislocations at the grain boundaries is still relatively high. Equilibrium boundaries are found only after long-term annealing.

At a temperature of 923 K, the Ni_3Al nanostructured alloy showed superplastic behaviour (the appearance of the specimen of Ni3Al after testing is shown in Fig. 7.6). For comparison, the initial (prior to testing) specimen is also shown. The experimental results show that the nanocrystalline specimens are characterised by very high plasticity of tensile loading (several hundred percent to failure)

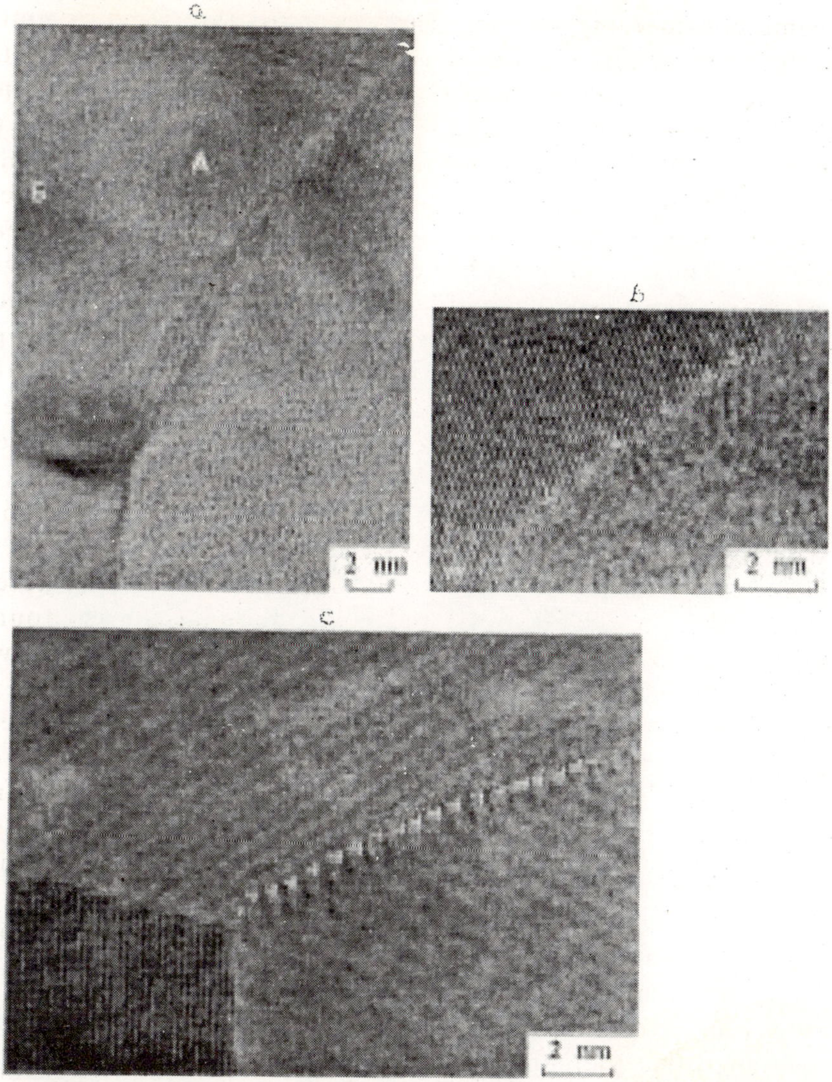

Fig. 7.5. High-resolution electron microscopy of specimens of nanocrystalline Ni_3Al alloy. *a, b*) after annealing at 650°C, 30 min, *c*) annealing at 1023 K, 3 min. Explanation in the text.

without formation of a visible macroscopic neck. This confirms the manifestation of the superplastic behaviour but in the nanocrystals, this behaviour has a number of unusual characteristics. In particular, the stress-strain (flow) curves for the nanostructured Ni_3Al at a strain rate of 10^{-3} s^{-1} and T = 923 K (Fig. 7.7*a*) are characterised by large areas of strain hardening at both temperatures with the maximum flow stresses reaching 0.9–1.5 GPa [35]. The resultant

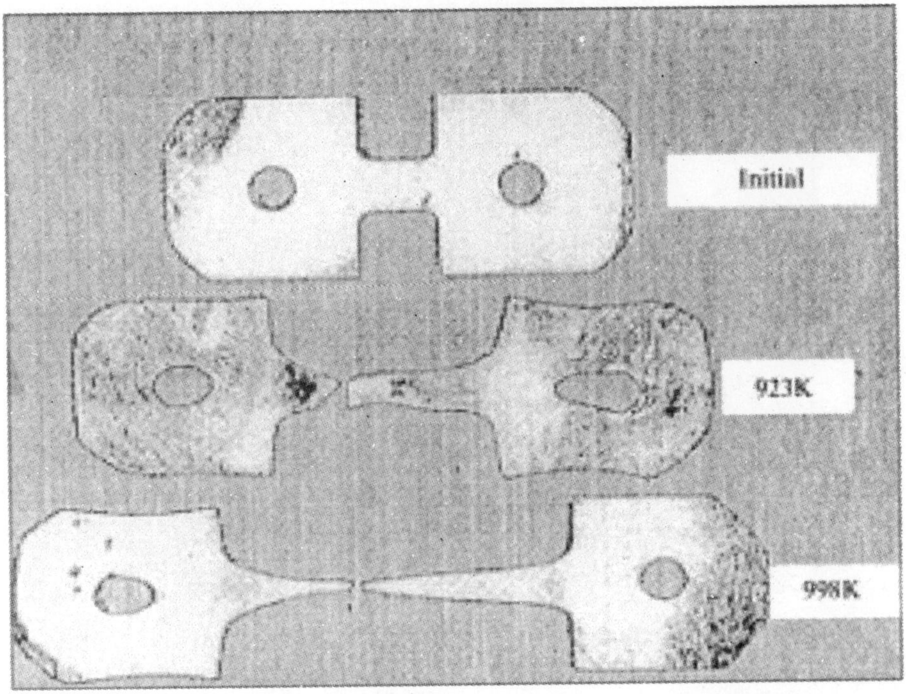

Fig. 7.6. Appearance of the nanostructured specimens of Ni_3Al prior to and after tensile loading at a rate of 10^{-3} s^{-1} at T = 923 K (elongation 390%) and 998 K (elongation 560%).

values are very high for the Ni_3Al intermetallic compound. The unusual effect of heat treatment on the behaviour of the alloy in tensile loading was also reported. For example, annealing for 1 min at a temperature of 1023 K in argon prior to tensile testing resulted in the situation in which the specimen did not show any superplastic behaviour even in the absence of grain growth. It should be mentioned that the superplasticity in the Ni_3Al nanostructured alloy was found at temperatures lower by more than 673 K than in the case of microcrystalline Ni_3Al [346]. Recently, it was attempted for the first time to determine the equation of state for superplastic nanocrystalline Ni_3Al [347]. To determine the rate sensitivity parameter, experiments were carried out with switching of the strain rate at a test temperature of 923 K. Figure 7.7 shows the flow curves (stress–strain) in the experiments carried out to determine the coefficient of rate sensitivity, and one of deformation curves at a constant tensile loading rate. This curve is typical of the specimens of nanocrystalline Ni_3Al. The curve contains a relatively long region characterised by hardening of the specimens at an almost constant rate. This is one of the special features of

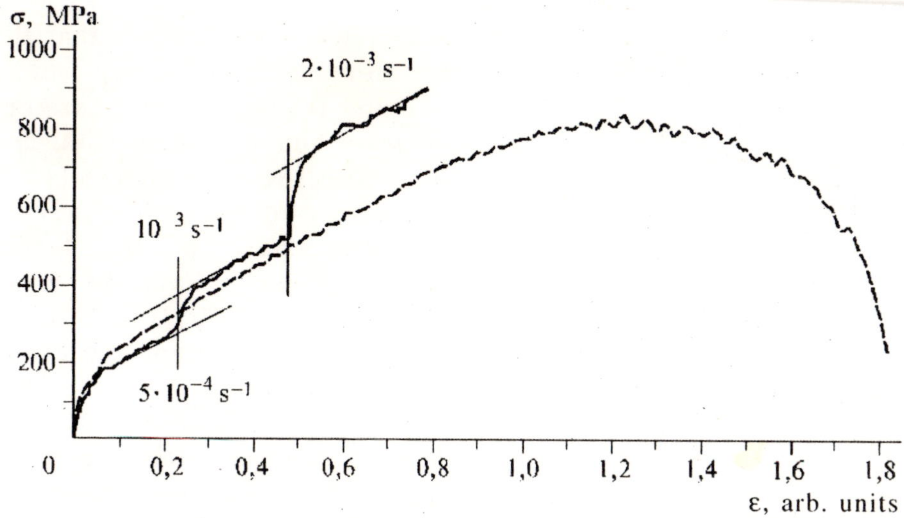

Fig. 7.7. Stress–strain curves for Ni_3Al (T = 923 K) with switching of the rates.

superplastic behaviour of the nanocrystalline materials. Figure 7.8 shows the dependence of the mean grain size as a function of the annealing time of nanocrystalline Ni_3Al at an annealing temperature of 923 K taking into account the fact that the process of tensile loading of the specimens to fracture lasts approximately 20 min, it may be concluded that the microstructure is stable at this temperature. No significant grain growth was detected: the scatter of the grain sizes, produced by the secant method, was 85–120 nm. Constructing the dependence of ln (n) on ln (d) and using the value of n = 2.6 obtained in the experiments with different strain rates, gives the exponent in the dependence of the flow stress on the mean grain size in the range 1.8–2.4. The determination of this parameter is important for understanding the nature of superplasticity in the nanostructured material. One of the special features of the superplasticity equation is that it describes the process in the steady flow stage, i.e., on the plateau. However, this plateau is not detected in the experiments with the specimens of Ni_3Al with a section with hardening. It is therefore necessary to carry out investigations of other nanostructured materials, directed to determination of the parameters of the equation of state.

The results of experiments with annealing of the nanocrystalline Ni_3Al and differential scanning calorimetry show that the temperature resulting in the occurrence of superplasticity is close to the temperature of the start of grain growth. Evidently, this is

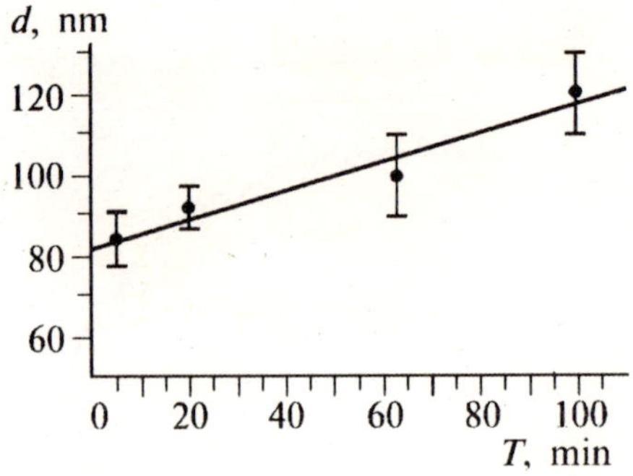

Fig. 7.8. Dependence of the mean grain size in isothermal annealing at T = 923 K on annealing time.

associated with the fact that the activation energy of superplasticity is equal to the activation energy of grain boundary diffusion, and the nanostructured materials are characterised by higher driving forces because of the larger volume fraction of the grain boundaries. Therefore, it is interesting to examine the effect of grain growth on the flow stress in the nanostructured materials. The superplasticity equation shows that grain growth results in hardening $\sigma \sim d^{p/n}$. However, the hardening, recorded in the superplasticity of the nanostructured specimens, can be explained only by grain growth. According to the experimental data, published in [37], $p/n \sim 1$, the mean grain size increased from 85 to 100 nm at T = 923 K, and the flow stress increased four times.

To explain the mechanism of the unusual superplastic behaviour of nanostructured Ni_3Al, special investigations were carried out by transmission electron microscopy, including high-resolution microscopy. This was carried out on foils prepared from a section of the working part of the specimen of Ni_3Al, deformed at a temperature of 923 K, strain 300% (Fig. 7.9) [348]. Although deformation was accompanied by some grain growth, the grain size remained smaller than 100 nm (Fig. 7.9*a*). The grains were not elongated and there was no indication of dislocation activity inside the grains, although detailed investigations were carried out by high-resolution electron microscopy of the boundaries of annealing twins (Fig. 7.9*b*) which formed in some grains of the alloy subjected to severe plastic deformation during heating prior to deformation. Similar twin boundaries are known by their capacity to easily trap and retain lattice dislocations [206, 340]. However, it was not possible to find dislocations in these internal interfaces. These experiments are in agreement with the theoretical calculations, demonstrating the difficulties of accommodation of grain boundary sliding by generation of lattice dislocations in nanograins during superplastic deformation [349]. Evidently, the superplastic deformation of the nanomaterials takes place as a result of grain boundary sliding and some diffusion accommodation without visible dislocation activity in the grains.

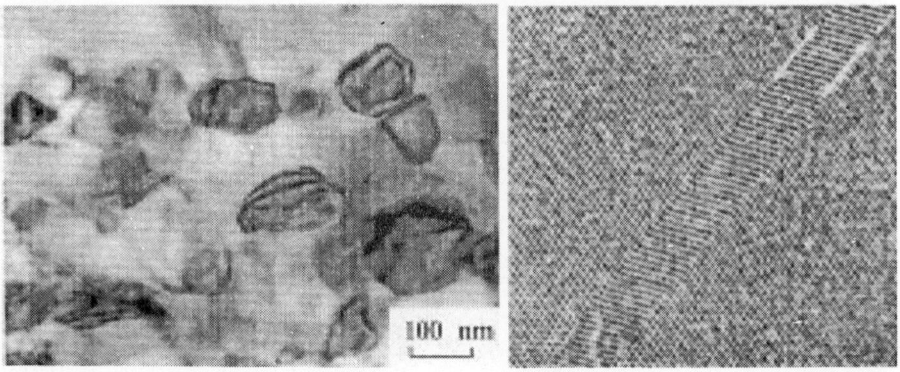

Fig. 7.9. Photographs of the nanostructured Ni_3Al. *a*) bright-field electron microscopic image, *b*) high-resolution photograph of the twin boundaries.

Attention will now be given to the process of superplastic flow in nanostructured aluminium alloy 1420, produced by torsion under pressure. The photographs of the microstructure of the alloy were produced by transmission electron microscopy (Fig. 7.10). The mean grain size in the specimen produced by severe plastic deformation in torsion was smaller than 100 nm. The diffraction pattern shown in the insert corresponds to the nanocrystalline condition. The grains are mainly equiaxed. After heating to 573 K there was some grain growth but the mean grain size did not exceed 200 nm.

The specimens prior to and after testing at different temperatures are shown in Fig. 7.11. The latter are typical of elongation without necking which is a characteristic feature of

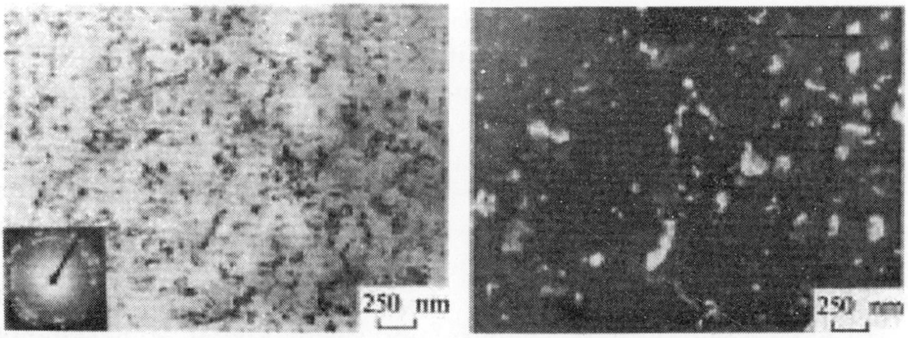

Fig. 7.10. Microstructre of 1420 alloy after torsion under pressure. *a*) bright-field image and the microdiffraction patent, *b*) dark field image.

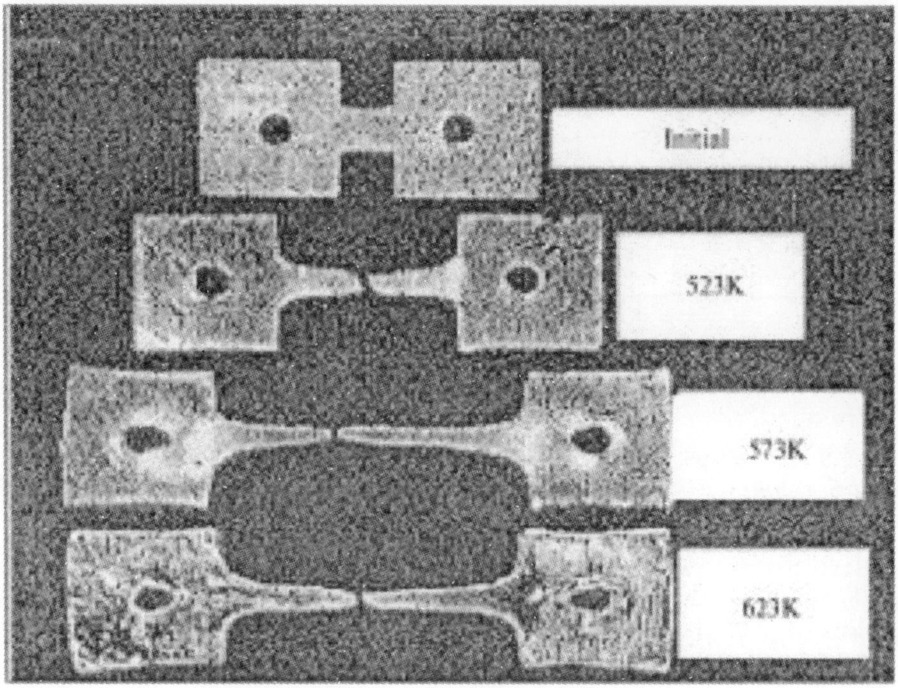

Fig. 7.11. Appearance of specimens of 1420 alloy after tensile loading.

superplastic behaviour. The second very important feature is the shape of the deformation curves characterised by considerable strain hardening at all strain rates, and the strain hardening was higher when the strain rate was higher. Nevertheless, this alloy shows superplasticity at high strain rates of up to $5 \cdot 10^{-1}$ s^{-1} and a relatively low temperature, 573 K [350].

Superplasticity of nanocrystalline nickel

As already mentioned, superplasticity in pure metals is a relatively rare phenomenon, because of grain growth. However, as in the case of pure nickel [213, 352], grain growth may support accommodation of grain boundary sliding by 'sweeping up' of triple joints. We believe that grain growth may play this positive role only up to the moment when the mean grain size is smaller than some critical value. We shall discuss examples of the experimental examination of superplastic deformation in pure nickel with a nanocrystalline structure.

In early studies of the superplasticity of nickel, experiments were carried out with uniaxial tensile loading [351] to examine the effect of fine grains (~4 μm) on the superplastic behaviour of nickel.

For comparison, experiments were carried out with coarse-crystalline nickel, produced by annealing at a temperature of 1173 K for 1 h. The mean grain size after this type of treatment was approximately 100 μm. The initial structure has the strongest effect at elevated temperatures. The results show that at T = 823 K the flow stress of fine-grained nickel is higher in comparison with coarse-crystalline nickel; however, a further increase of deformation temperature of nickel is characterised by the opposite situation – coarse-grained nickel is stronger. In fine-grained nickel, plasticity increases with increasing temperature, reaching the maximum value (180°) at T = 1073 K. Figure 7.12 shows the 'stress–strain' curves, recorded in [351] at a temperature of 1073 K. Fine-grained nickel shows the typical diagram of superplastic flow. In the case of nickel with larger grains, the behaviour of the specimens in deformation is different, i.e. hardening takes place up to fracture. The initial microstructure also has a strong effect on the rate dependence of the flow stress and relative elongation (Fig. 7.13). In the case of low strain rates, the flow stress for fine-grained nickel is lower than for coarse-grained nickel. The situation changes for rates higher than 5.10^{-2} s^{-1}. For the strain rate of $5{\cdot}10^{-3}$ the coefficient of rate sensitivity for fine-grained nickel is 0.38, for coarse-grained nickel it is 0.22 (Fig. 7.14). The plasticity of coarse-grained nickel monotonically increases with increasing strain rate, and that of fine-grained nickel has a maximum at $5{\cdot}10^{-2}$ s^{-1} and then rapidly decreases (Fig. 7.13).

A recent detailed examination of the properties of nickel with a nanocrystalline structure is very interesting. These investigations have been carried out on specimens of nickel produced by

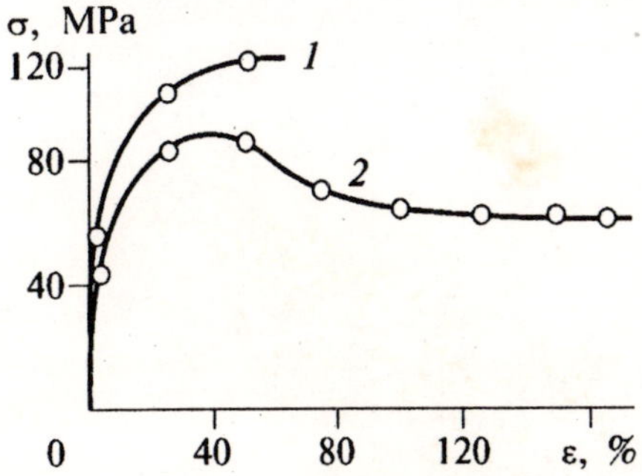

Fig. 7.12. True stress–strain curves for nickel. 1) d_0 = 100, 2) 4 μm, T = 1073 K, = $5{\cdot}10^{-3}$ s^{-1}.

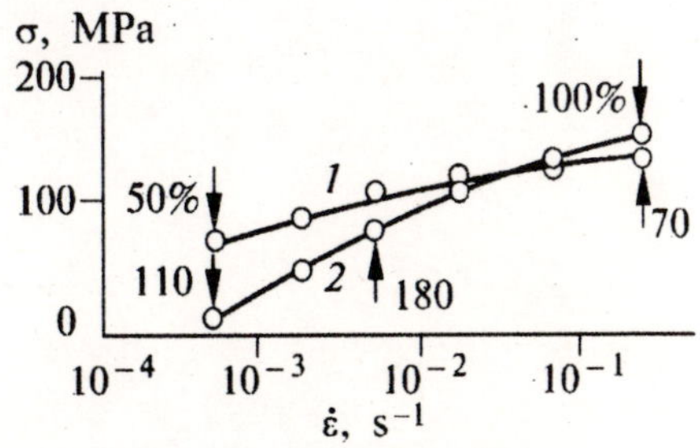

Fig. 7.13. Dependence of flow stress (σ) on strain rate for nickel. *1*) d_0 = 100, 2) 4 μm. T = 1073 K. The numbers on the graph indicate the value of the maximum elongation, obtained for the given strain rate.

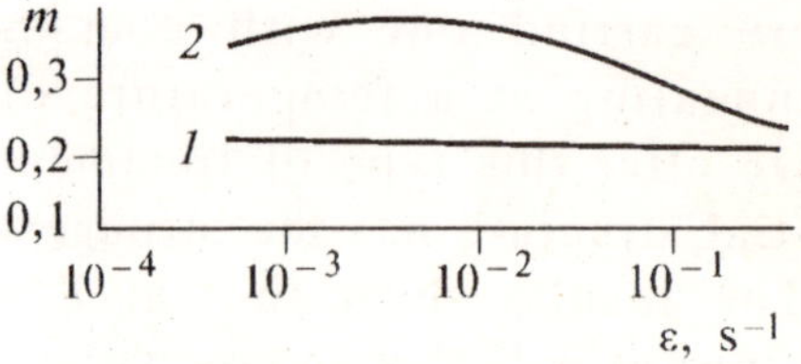

Fig. 7.14. Dependence of the coefficient of rate sensitivity (m) of strain rate for nickel. *1*) d_0 = 100, *d*) 4 μm, T = 1073 K.

electrodeposition [42, 43]. The bright field image of the microstructure and microdiffraction patterns are presented in Fig. 7.15. The mean grain size of the initial structure of nickel is approximately 35 nm. The mean grain size, produced on the basis of the dark field image, coincided with the results of x-ray measurements. Experiments with uniaxial tensile loading were conducted in a mini-machine (Fig. 7.16). The external appearance of mini-specimens prior to and after tensile loading is shown in Fig. 7.17. Elongation was measured directly of the gauge length of the specimen and not on the basis of the displacement of the clamps. The form of the working part demonstrates uniform elongation without the formation of a localised neck. The maximum elongation was recorded at a temperature of 693 K and a strain rate of$1\cdot10^{-3}$ s^{-1}. The stress–strain curves, recorded at a rate of $1\cdot10^{-3}$ s^{-1}, for three different test temperatures (552, 623 and 693 K) are shown in Fig. 7.18. The increase of temperature above 553 K results in a rapid decrease of the flow stress and already at T = 623 K the specimens of nanocrystalline nickel show plasticity greater than 200%. A further increase of temperature increases the plasticity of the specimens,

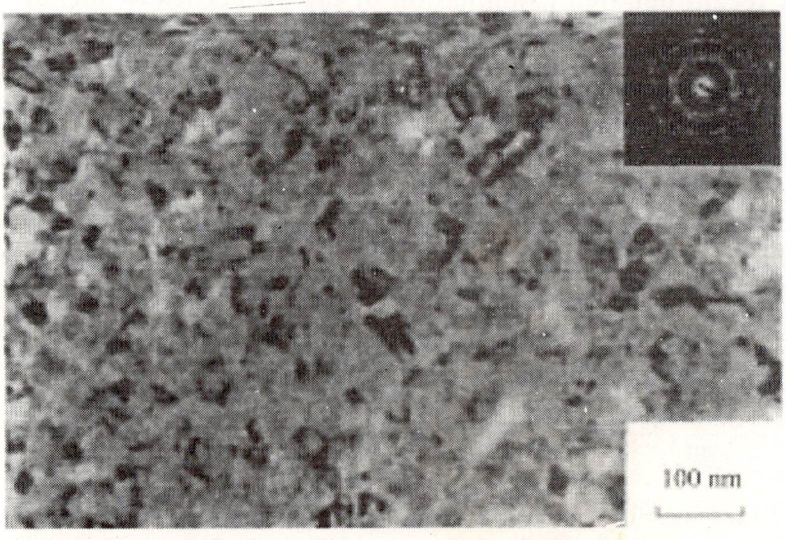

Fig. 7.15. Bright-field image and the microdiffraction pattern of electrodeposited nickel.

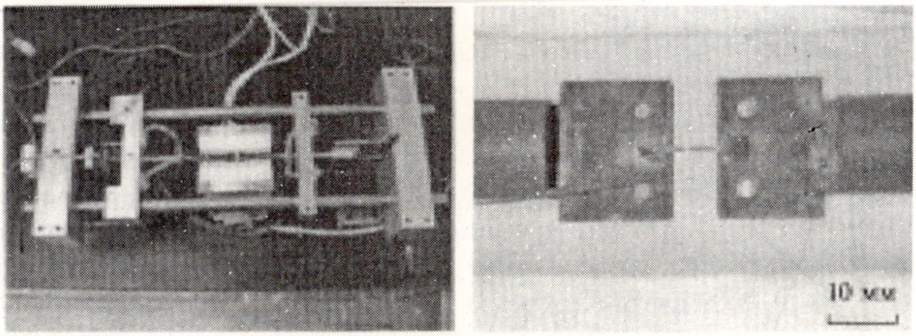

Fig. 7.16. The general view of the testing machine and the clamps used for tensile loading of minispecimens. The machine was designed and constructed in the laboratory of Professor Mukerjee (Davies, USA) with the support of the National Scientific Fund of the USA.

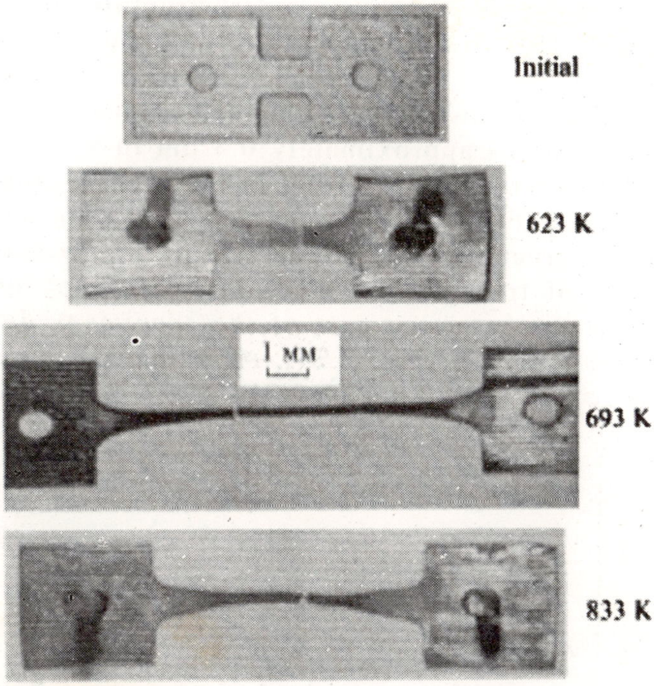

Fig. 7.17. External appearance and form of the specimens of nanocrystalline nickel prior to and after tensile testing.

with the maximum value recorded at a temperature of 693 K, equalling 895%, which is the highest recorded elongation value for pure nickel. The transition of the nickel specimens from the condition with low plasticity to the condition with high plasticity coincides with the exothermic peak on the curve recorded in a

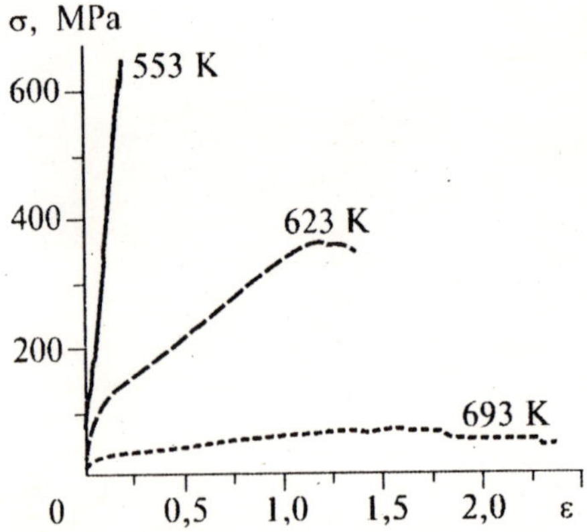

Fig. 7.18. True flow stress-strain curves for the strain rate of $1 \cdot 10^{-3}$ s^{-1}.

differential scanning calorimeter (DSC). Figure 7.19 shows a signal from the DSC at a heating rate of the specimens of 40 deg/min and the elongation values, recorded at the given deformation temperature. On the basis of the temperature profile of the mechanical tests experiments were carried out with annealing of nanocrystalline nickel in order to determine the microstructure at the moment of the start of tensile loading. Figure 7.20*a* shows the microstructure of the specimen after heating from room temperature to a temperature of 623 K with the same rate as in the mechanical tests (approximately 40–50°/min). The microstructure remained nanocrystalline with the precipitated grains approximately 0.3 μm in size. In heating to 693 K (the experiments resulting in the highest elongation), the microstructure (Fig. 7.20*b*) is characterised by a more homogeneous grain structure with a mean size of approximately 0.5 μm. The results of transmission electron microscopy of thin foils, produced from the gauge part of the specimens deformed at a temperature of 623 K (Fig. 7.21), show that the mean grain size in the gauge part was 1.3 μm along the tensile loading axis and 0.6–

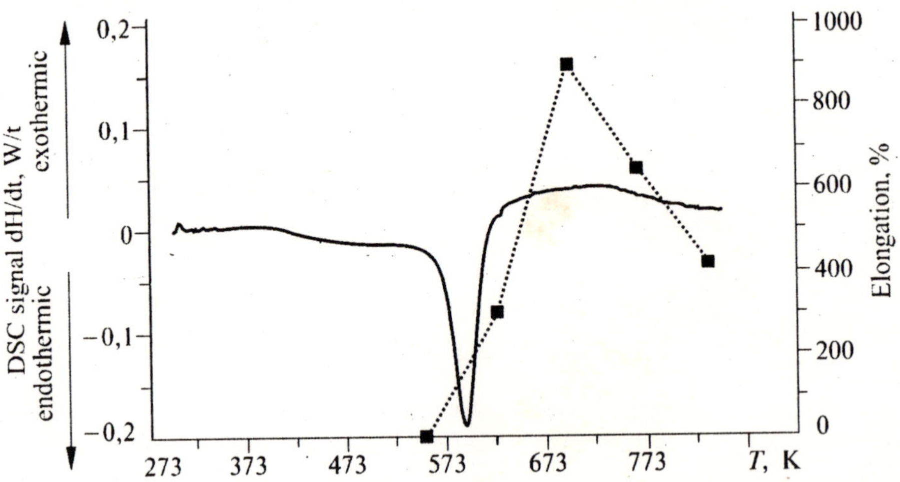

Fig. 7.19. The curves of differential scanning calorimetry (solid line) combined with the dependence of maximum elongation (broken line) on test temperature.

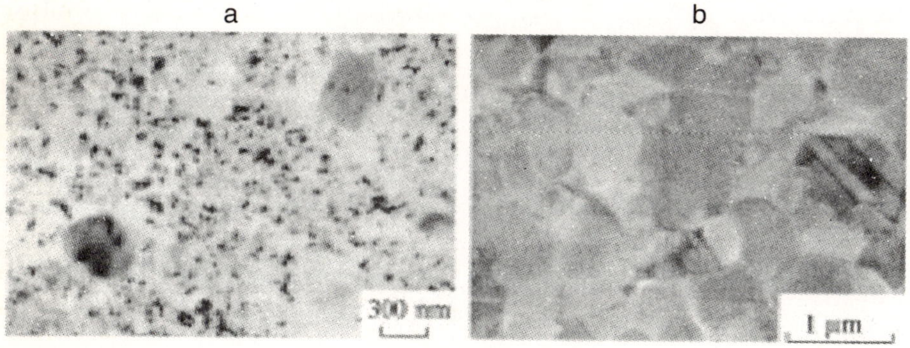

Fig. 7.20. The microstructure of nickel specimens after heating to 623 K (*a*) and 693 K (*b*) with the heating rate of 40–50 deg/min.

0.7 μm in the cross section. The mean grain size of the specimens, deformed at 693 K, was approximately 2 μm along the tensile loading axis and 1 μm in the cross section. Thus, in heating to the temperature of the start of deformation the grains grew very rapidly, i.e. the mean size increased approximately by three orders of magnitude: from 35 nm to 0.3–0.5 μm. Nevertheless, the structure of nickel just prior to the start of testing remained ultrafine-grained, considerably finer than that recorded in the experiments in [351]. As already mentioned, the most important characteristic of the process of superplastic deformation is the exponent (n) in equation (7.9) which may be measured in the experiments with switching of

Fig. 7.21. The microstructure of nickel after superplastic deformation at 693 K in the gauge part of the specimen.

the rate of superplastic deformation. The values of n can be calculated from the equation [352]:

$$n = \frac{\ln(\dot{\varepsilon}_2/\dot{\varepsilon}_1)}{\ln(\sigma_2/\sigma_1)}. \tag{7.10}$$

Such experiments were carried out in [352] and the results are shown in Fig. 7.22 for two test temperatures: 623 and 693 K. The strain rate was changed in jumps: from $1 \cdot 10^{-3}$ s^{-1} to $2 \cdot 10^{-3}$ s^{-1} for T = 623 K and from $1 \cdot 10^{-3}$ s^{-1} to $5 \cdot 10^{-2}$ s^{-1} for 693 K. In both cases, the rate sensitivity exponent decreases with increasing strain. These experimental results are unusual because until recently it was not possible to obtain such high elongation values in pure metals because of grain growth at the test temperature. The transition of nanocrystalline nickel from the condition with low plasticity to the condition with high plasticity, detected at a test temperature higher than 553 K, corresponds to the transition from the purely dislocation mechanism of the formation to the thermally activated mechanism of grain boundary sliding. This behaviour is characteristic of nickel [359], but the transition temperature is considerably lower and the jump in the flow stress is much greater. This is the result of the small grain size: for low temperatures (and, consequently, the dislocation mechanism of deformation) the high flow stress is caused by the Hall–Petch effect, and at elevated temperatures the decrease of the flow stress is determined by the small grain size, as indicated by the superplasticity equation (7.9). In the materials with small grains in transition to the thermally activated process of deformation (grain boundary sliding and diffusion and dislocation accommodation) the thermal energy is

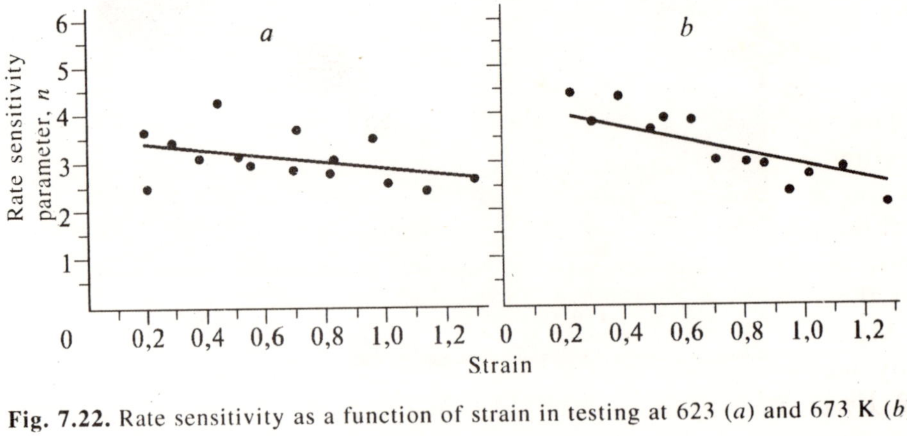

Fig. 7.22. Rate sensitivity as a function of strain in testing at 623 (a) and 673 K (b).

redistributed between the processes of superplastic flow and grain growth. Both processes are characterised by approximately the same activation energy, equal to the activation energy of grain boundary diffusion. The phenomenon of grain growth in superplastic deformation has been known for some time but has not been investigated for the materials with the ultrafine-grained structure. It appears that as soon as the driving force of grain growth greatly decreases with an increase of the mean grain size to approximately 1 μm, the mechanisms of superplastic behaviour start to operate and grain boundary sliding takes place together with grain growth. The latter slows down because part of the thermal energy is used for the realisation of superplastic deformation. The rate sensitivity parameter is an indicator of the controlling mechanism. For both deformation temperatures, the rate sensitivity parameter has a tendency to decrease from 4 to 2 with an increase of strain. The coefficient of rate sensitivity *m* (the value reciprocal to *n*) increases from 0.25 to 0.5. This is in agreement with the results published in [351]: $n \approx 2.5$ for fine-grained nickel deformed at a temperature of 1073 K at a strain rate of $5 \cdot 10^{-3}$ s^{-1}. The variation of the rate sensitivity parameter with increasing strain indicates the evolution of the microstructure which takes place in parallel with the superplastic flow.

Superplasticity of nanostructured alloys

Recent investigations show that the application of equal-channel angular pressing not only greatly reduces the temperature of superplastic flow but also results in high-rate superplasticity in ultrafine-grained alloys with the grain size of 0.5–1.0 μm [353, 354]. It has also been established [355] that the presence of preferentially high-angle grain boundaries is one of the most important requirements for achieving high-rate superplasticity in industrial alloys.

The decrease of the temperature of eqtual-channel angular pressing of aluminium alloy 1420 resulted in the mean grain size of 0.4 μm and the formation of the two-phase structure, containing circular particles of a second phase with the size of 0.1–0.2 μm [356]. Analysis of the results of energy-dispersion analysis shows that these particles are T-Al_2LiMg phase, which forms in the Al-Li–Mg alloy at temperatures above 473 K. After this treatment, the alloy was characterised by high stability up to a temperature of 673 K and the grain size during heating did not exceed 1 μm,

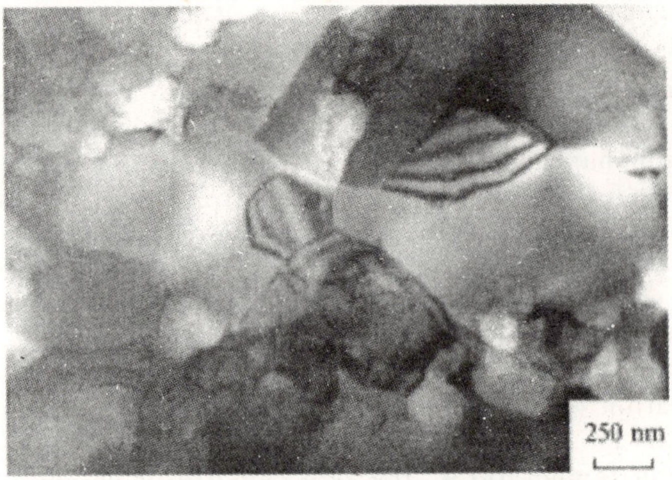

Fig. 7.23. The microstructure of 1420 alloy after equal-channel angular pressing and heating to 673 K.

although some relaxation of the structure took place, associated with the decrease of the level of internal stresses (Fig. 7.23). The alloy with this type of structure shows the highest high-rate superplastic properties. In particular, tests at a temperature of 673 K and a strain rate of 10^{-1} s^{-1} showed that the elongation to fracture was 1240% (Fig. 7.24). At the same time, the flow stress was low, less than 30 MPa. The alloy showed increased superplasticity even at a strain rate of 1 s^{-1}. In these conditions, the elongation to fracture was 1000%. Figure 7.24 shows the appearance of specimens indicating their uniform elongation without necking during deformation. These results show superplasticity at the highest strain rate, recorded in the literature for industrial alloys. It should be mentioned that the elongation to fracture higher than 1000% was also recorded in a sub-microcrystalline titanium alloy produced by multiple forging [58] but in this case the strain rate did not exceed 10^{-3} s^{-1}.

In the Al–4% Cu–0.5% Zr alloy after equal-channel angular pressing, the mean grain size was approximately 150 nm and high-dispersion particles of Al3Zr with the size of up to 30 nm were found [343]. The specimens of the Al–Cu–Zr alloy were subsequently subjected to tensile loading at a temperature of 523 K with different strain rates in the range from $2.8 \cdot 10^{-5}$ to $1.4 \cdot 10^{-3}$ s^{-1}. The results show that the alloy shows very high elongation to fracture values, regardless of the relatively low test temperature. The maximum elongation was 850% at the initial strain

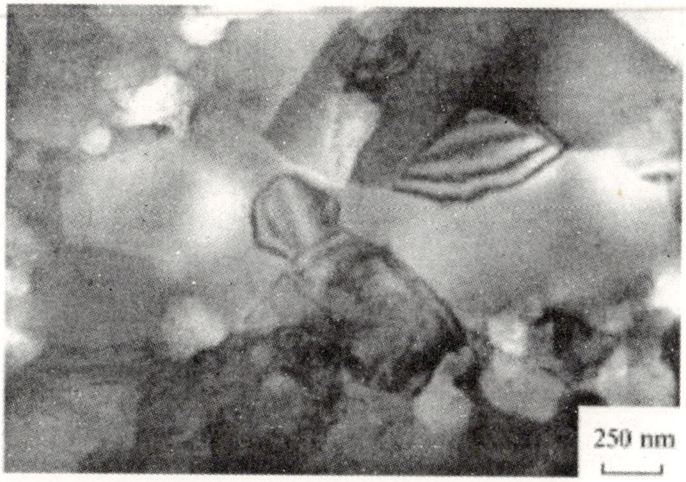

Fig. 7.24. Appearance of the specimens of 1420 alloy, produced by equal-channel and after tensile loading at 673 K at a rate of 10^{-1} s^{-1}.

rate of $1.4 \cdot 10^{-4}$ s^{-1}. The coefficient of rate sensitivity m for this case was equal to 0.46. For comparison, the same alloy with the grain size of 8 μm shows a similar superplastic behaviour only at a temperature of 773 K [282].

The Zn–22% Al eutectoid alloy is a classic two-phase superplastic alloy showing the elongation in tensile testing greater than 2000% in the optimum temperature–rate conditions of deformation (temperature 523 K, strain rate 10^{-3} s^{-1}) [283]. Usually, superplasticity in this alloy is obtained at the grain size in the range 1–5 μm. In order to examine the effect of the nanostructure on the superplastic behaviour, spe-cimens of the alloy were subjected to two different schemes of severe plastic deformation: equal-channel angular pressing and deformation by torsion.

The Zn–22% Al alloy, subjected to equal-channel angular pressing, have the mean grain size of approximately 0.5 μm and showed high superplastic properties at very high strain rates [356]. As indicated by Table 7.2, at a strain rate of the order of $3.3 \cdot 10^{-1}$ s^{-1} the alloy, subjected to equal-channel angular pressing, showed very high elongation values, up to 1540%, whereas in the alloy with the micrograin structure, the maximum elongation was recorded at a strain rate of 10^{-3} s^{-1} [283]. To obtain the smallest grains, the speci-mens of the Zn–22% Al alloy were also quenched followed by deformation by torsion. This procedure resulted in the formation of a two-phase nanoduplex structure with the mean grain size of approximately 80 nm (Fig. 7.25) [357]. At the same time,

Table 7.2. Maximum values of the elongation of the Zn–22% Al alloy subjected to equal-channel angular pressing, after tensile loading at different test rates

Test temperature, K	Elongation, %			
	1	2	3	4
373	450	–	–	–
423	700	940	–	–
473	–	–	1900	1540

energy-dispersion analysis showed changes in the chemical composition of both phases. For example, it was found that the zinc content of the Al-phase region was 10% which is approximately 5 times higher than in the equilibrium condition. The superplastic behaviour of the specimens was recorded at a temperature of 393 K at a strain rate of 10^{-4} s^{-1}. Nevertheless, the elongation to fracture was relatively small, 280%. For comparison: the elongation of the same alloy with the mean grain size of 0.5 μm (produced by equal-channel angular pressing) in testing in the same temperature-rate conditions reached 600%.

Thus, the nanostructured materials showed: superplasticity at temperatures considerably lower in comparison with microcrystalline alloys, and the possibility of realisation of superplasticity at very high strain rates. The nature of these effects was discussed in [213] where it was shown that the non-equilibrium condition of the grain boundaries in the nanostructured materials, leading to the acceleration of the dynamic processes at the grain boundaries, may result in a considerable decrease of the temperature of superplastic deformation. In addition to this, grain boundaries, distorted by the dislocations, may also be responsible for the expected manifestation of high-rate superplasticity in the nanostructured materials because of acceleration of grain boundaries sliding at these boundaries [358].

At the same time, the effect of the mean grain size on the superplastic behaviour of the nanostructured alloys, as shown by the previously mentioned experimental results, is not unambiguous. For example, the specimens of the Zn–22% Al alloy, subjected to equal-channel angular pressing with the mean grain size of approximately 0.5 μm, are characterised by high superplastic properties. At the same time, the level of the superplastic properties of this alloy in the nanocrystalline condition with the mean grain size of 80 nm was considerably lower. The identical tendency was found in the Ni_3Al (Cr) alloy, alloyed with B, in which the superplastic behaviour in the nqnocrystalline condition was observed at a temperature of 923 K which is considerably lower in comparison with the fine-grained

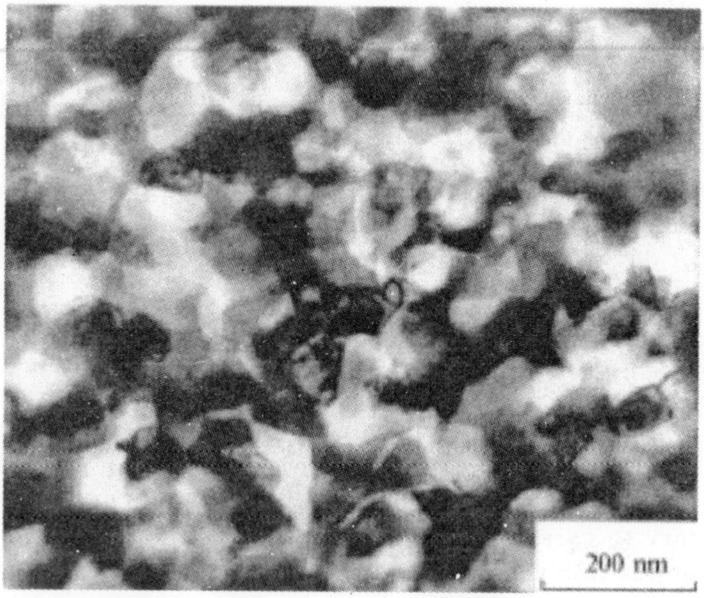

Fig. 7.25. The microstructure of the Zn–22% Al two-phase alloy subjected to severe plastic deformation by torsion at room temperature.

specimen (1423 K) [346]. Nevertheless, the elongation to fracture is only approximately 400% at the flow stress of 1500 MPa.

The results indicate that the decrease of the mean grain size to 100–300 nm increases the superplastic properties, but the role of the further refining of the grains in the manifestation of the superplasticity effect evidently requires more detailed investigations. In particular, in [348] it was reported that the generation of dislocations in the nanograins may be complicated and this results in the deceleration of the accommodation–dislocation processes. In this connection, one could expect a a decrease of the level of the superplastic properties of the nanocrystalline materials with the grain size smaller than some critical value. At the same, in the nanocrystalline Zn–22% Al alloy the chemical composition of the Zn and Al phases changed which may have also resulted in a decrease of the superplastic properties. It is important to note the atomic disordering in the $Ni_3Al(Cr)$ alloy, subjected to severe plastic deformation; this may have also influenced the superplastic behaviour. Consequently, the problem of the possibility of obtaining higher superplastic properties in the investigated alloys with a further decrease of the grain size remains open.

Thus, the application of the methods of equal-channel angular pressing for the production of sub-micrograin structures has resulted in improved superplastic properties, namely the low-temperature and high-rate superplasticity in a number of alloys. The high superplastic properties, detected in the nanostructured alloys, depend not only on the grain size but also on the type of resultant microstructure. The nanocrystalline alloys show a very unusual superplastic behaviour, associated with marked strain hardening which is evidently determined by the variation of the deformation mechanisms as a result of difficulties in the dislocation accommodation of grain boundary sliding in the fine grains [257].

The detection of superplasticity in the ultrafine-grained materials at relatively low temperatures and very high strain race indicates the possibility of a large increase of the efficiency of practical application of superplastic forming in different industrial alloys as a result of the refining of the structure of the alloys. However, to obtain high superplastic properties in the ultrafine-grained alloys, it is necessary to carry out detailed control of the microstructure and phase composition of the alloys.

References

1. Gleiter H., Progress Mater. Sci., 1989 VOl.33, No.4, 223.
2. Morokhov I.D., Trusov L.D., Lapovok V.I., Physical phenomena in ultrafine media, Moscow, Nauka, 1984.
3. Flagan R. C., In: Proc. of the NATO ASI on Nanostructructured Materials: Science & Technology. Dordrecht-Boston-London: Kluwer Acad. Publ., 1998, Vol. 50, 15.
4. Chow G. M., In: Proc. of the NATO ASI on Nanostructured Materials: Science & Technology. Dordrecht-Boston-London: Kluwer Acad. Publ., 1998, Vol. 50, 31.
5. Koch C.C., Cho Y.S., Nanostructured Materials, 1992, Vol. 1, 207.
6. Morris D. G., Mechanical behaviour of nanostructured materials. Switzerland: Trans. Tech. Publication Ltd., 1998, 85.
7. Ultrafine-grained materials prepared by severe plastic deformation (ed. by R.Z. Valiev), Annales de Chimie. Science des Materiaux, 1996, Vol. 21, No. 6-7, 369.
8. Valiev R. Z., Korznikov A. V., Mulyukov R. R., Mater. Sci. Eng., 1993, Vol.A186, 141.
9. Langford G., Cohen M., Trans. ASM., 1969, Vol.82, 623.
10. Pavlov V.A., FMM, 1989, Vol.6, 924.
11. Rybin V.V., High plastic deformation and failure of metals, Moscow, Metallurgiya, 986.
12. Valiev R. Z., Nanostructured Materials, 1995, Vol.6, 73.
13. Gleiter H., Marquardt P., Zs Metallkunde, 1984, Vol.75, No. 4, 263.
14. Birringer R., Herr U., Gleiter H., Trans. Japan. Inst. Met. Suppl., 1986, Vol. 27, 43.
15. Siegel R. W., J. Phys. and Chem. Solids, 1994, Vol. 55, No. 10, 1097.
16. Andrievskii R.A., Powder Material Science, Moscow, Metallurgiya, 1991.
17. Matthews M. D., Pechenik A., J. Amer. Ceram. Soc., 1991, Vol. 74, No. 7, 1547.
18. Chen D.-J., Maya M. J., Nanostructured Materials, 1992, Vol. 2, No. 3, 469.
19. Andrievski R. A., Kalinnikov G. V., Potafeev A. F. et al, Nanostructured Materials, 1995, Vol. 6, No. 1– 4, 353.
20. Rabe Ò., Wasche R., Nanostructured Materials, 1995, Vol.6, No.1–4, 357.
21. Ivanov V.V., Yavorskii N.A., Kotov Yu.A., et al, Dokl. AN SSSR, 1984, Vol. 275, No. 4, 873.
22. Ivanov V.V., Paranin S.N., Gavrilin E.A., et al, SFKhT, 1992, Vol.5, No.6, 1112.
23. Avakumov E.G., Mechanical methods of activation of chemcial processes, Novosibirsk, Nauka, Siberian Division, 1988.
24. Avvakumov E.G. (ed), Mechanical synthesis in inorganic chemicstry, Novosibirsk, Nauka, Siberian Division, 1991.
25. Mechanical Alloying / Ed. P. H. Shingu. Switzerland: Trans. Tech. Publications, 1992. – 840 p.
26. Gaffet et al., Mater. Trans., 1995, Vol. 36(2), 1980.
27. Yavari A. R., Desre P. J., Benameur Ò., Phys. Rev. Lett., 1992, Vol. 68, No. 14, 2235.
28. Fecht H.-J., Nanostructured Materials, 1995, Vol. 6, No. 1-4, 33.
29. Bologh J., Bujaso L., Faigel G. et al, Nanostructured Materials, 1993, Vol.

2, No. 1, 11.
30. Mechanical behavior of nanostructured materials, Morris D. G. Switzerland: Trans. Tech. Publications, 1998, 100 p.
31. Salishchev G.A., Valiakhmetov O.R., Galeev R.M., Malysheva S.P., Metally, 1996, No.4, 86.
32. Chakkingal U., Suriadi A.B., Thomson P.F., Scripta Mater., 1998, Vol.39, 677.
33. Mishin O.V., Alexandrov I.V., Golubev O.V., Greshnov V.M., Valiev R.Z., In: Proc. of the Intern. Simposium "Metallography'95". Stara Lesna, Slovakia, 1995.
34. Zhorin V.A., Shashkin D.P., Enikolopyan N.S., et al. Dokl. AN SSSR, 1984, Vol. 278, 144.
35. Kuznetsov R.I., Bykov V.I., Chernyshev V.P., et al, Plastic deformation of solids and pressure, Institute of Physics of Metals, Ural Scientific Centre of the Russian Academy of Sciences, Sverdlovsk, 1982.
36. Bridgman P.V., Examination of high plastic strains and fracture, Moscow, IL, 1955.
37. Smirnova N.A., Levit V.I., Pilyugin V.I., et al, FMM, 1986, Vol. 61, 1170.
38. Valiev R.Z., Ivanisenko Yu.V., Rauch E.F., Baudelet B., Acta Mater. 1997, Vol.44, 4705.
39. Valiev R.Z., In: Synthesis and processing of nanocrystalline powder (ed. by David L. Bourell). The Minerals, Metals and Materials Society, 1996, 153.
40. Mishra R.S., Valiev R.Z., Mukherjee A.K., Mater. Sci. Forum, 1996, Vol.225-227, 605.
41. Segal V.M., Reznikov V.I., Drobyshevskii A.E., Kopylov V.I., Izv. AN SSSR. Metally, 1981, No.1, 115.
42. Segal V.M., Reznikov V.I., Kopylov V.I., et al, Processes of plastic structure formation of metals, Minsk, Navuka i Tekhnika, 1994.
43. Akhmadeev N.A., Valiev R.E., Kopylov V.I., Mulyukov R.R., Izv. RAN. Metally, 1992, No.5, 96.
44. Valiev R.Z., Tsenev N.K., In: Hot deformation of aluminum alloys (ed. by T.G. Langdon, H.D. Merchant, J.G. Morris, M.A. Zaidi). TMS. Warrendale, PA., 1991, 319.
45. Valiev R.Z., Krasilnikov N.A., Tsenev N.K., Mater. Sci. Enginer., 1991, Vol.A137, P.35.
46. Ferrase S., Segal V.M., Hartwig K.T., Goforth R.E., Metall. Mater. Trans., 1997, Vol.28A, 1047.
47. Iwahashi Y., Horita Z., Nemoto M., Langdon T.G., Acta Mater, 1997, Vol.45, 4733.
48. Iwahashi Y., Furukawa M., Horita Z., Nemoto M., Langdon T.G., Met. Trans. A., 1998, Vol.29A, 2245.
49. Segal V. M., Mater. Sci. Engin., 1995, Vol.A197, 157.
50. Iwahashi Y., Horita Z., Nemoto M., Langdon T.G., Met. Trans. A., 1998, Vol.29A, 2503.
51. Iwahashi Y., Horita Z., Nemoto M., Langdon T.G., Acta Mater., 1998, Vol.46, 1589.
52. Iwahashi Y., Wang J., Horita Z., Nemoto M., Langdon T.G., Scripta Mater., 1996, Vol.35, 143.
53. Iwahashi Y., Horita Z., Nemoto M., Langdon T.G., Acta Mater., 1998, Vol.46, 3317.

54. Furukawa M., Iwahashi Y., Horita Z., Nemoto M., Langdon T.G., Mater. Sci. Eng., 1998, Vol. A257, 328.
55. Valiakhmetov O.R., Galeev R.M., Salishchev G.A., FMM, 1990, No. 10, 204.
56. Galeev R.M., Valiakhmetov O.R., Salishchev G.A., Metally, 1990, No.4, 97.
57. Imayev R.M., Imayev V.M., Salishchev G.A., J. Mater. Sci., 1992, Vol.27, 4465.
58. Salishchev G.A., Valiakhmetov O.R., Galeyev R.M., J. Mater. Sci., 1993, Vol.28, 2898.
59. Kaibyshev O., Kaibyshev R., Salishchev G., Mater. Sci. Forum, 1993, Vol.113-115, 423.
60. Salischev G.A., Imaev R.M., Imaev V.M., Gabdulin N.K., Mater. Sci. Forum, 1993, Vol.113-115, 613.
61. Valitov V.A., Salishchev G.A., Mukhtarov Sh.Kh., Metally, 1994, No.3, 127.
62. Salishchev G.A., Valiakhmetov O.R., Valitov V.A., Mukhtarov S.K., Mater. Sci. Forum, 1994, V.170-172, 121.
63. Mazurskii M.I., Murzinova M.A., Salishchev G.A., Afonichev D.D., Metally, 1995, No.6, 83.
64. Gleiter H., Physica status solidi. B., 1992, Vol. 172, No 1, 41.
65. Harrison L.G., Trans. Faraday Soc., 1961, Vol.57, No 7, 1191.
66. Kaur I., Mishin Yu., Gust W. Fundamentals of Grain and Interphase Boundary Diffusion, 3rd ed, John Wiley & Sons Ltd, 1995.
67. Kaur I., Gust V., Diffusion in grain and phase boundaries, Moscow, Mashinostroenie, 1991.
68. Hart E.W., Acta met., 1957, Vol.5, No 10, 597.
69. Fischer J.C., J.Appl. Phys., 1951, Vol. 22, No 1, 74.
70. Bokshtein B.S., Magidson I.A., Svetlov I.L., FMM, 1958, No.6, 1040.
71. Gibbs G.B., Phys. Stat. Solidi, 1966, Vol.16, 27.
72. Whipple R.T.P., Philos. Mag., 1954, Vol.45, No. 351, 1225.
73. Suzuoka T., Trans. Jap. Inst. Metals, 1961, Vol.2, No.1, 25.
74. Mishin Yu.M., Razumovskii I.M., Mathematical models and method of determinination of diffusion parameters of individual boundaries, In: Structure and properties of internal interfaces in metals, Edited by B.S. Bokshtein, et al, Moscow, Nauka, 1988.
75. Levine H.S. and MacCallum C.J., J. Appl. Phys., 1960, Vol.31, 595.
76. Suzuoka T., J. Phys. Soc., Japan, 1964, Vol.19, 839.
77. Le Claire A.D., Brit. J. Appl. Phys., 1963, Vol.14, No.2, 351.
78. Mishin Yu.M., Poverkhnost'. Fizika, Khimiya, Mekhanika, 1983, No.6, 22.
79. Mishin Yu.M., Methods of determination of the parameters of boundary diffusion: theory and experimental verification, Author's abstract of the dissertation for the title of candidate of physico-mathematical sciences, Moscow, 1985.
80. Razumovskii I.M., Diffusion along internal interfaces and structural stability of creep-resisting alloys, Author's abstract of the dissertation for the title of candidate of physico-mathematical sciences, Moscow, 1988.
81. Mishin Yu.M., Razumovskii I.M., Poverkhnost'. Fizika, Khimiya, Mekhanika, 1983, No.7, 37.
82. King A.H., Intern. Materials Rev., 1987, Vol. 32, No 4, 173.
83. Hillert M., Purdy G.R., Acta metall., 1978, Vol. 26, No 2, 333.
84. Tu K.H., J. Appl. Phys., 1977, Vol. 48, No 8, 3400.

85. Den Broeder F.I.A., Acta metall., 1972, Vol. 20, No 3, 319.
86. Ievlev V.M., Trusov L.I., Kholmyanskii V.A., Structural transformation in films, Moscow, Metallurgiya, 1988.
87. Parthasarathy T.A., Shewmon P.G., Scripta metall., 1983, Vol. 17, 943.
88. Chen F.S., King A.H., Scripta metall., 1986, Vol. 20, No 10, 1401.
89. Y. Yamamoto, M. Moriyama, M. Kajihara and T. Mori., Acta mater., 1999, Vol. 47, No. 6, 1757.
90. Panin V.E., et al, Increasing the thermal stability of composite materials based on nickel hardened by tungsten and molybdenum fibres by directional alloying of the matrix, In: Structure and properties of creep-resisting metallic materials, Moscow, Nauka, 1973.
91. Geguzin Ya.E., Kivets V.I., FMM, 1973, Vol.36, 1043.
92. Geguzin Ya.E., Klinchuk Yu.I., FMM, 1974, Vol.37, No.5, 1099.
93. Hoffman J., Hofmann S., Tillman L., Z. Metallkunde, 1974, Vol.65, 721.
94. Grunling H., Hofer G., Z. Werkstofftechn., 1974, Vol.5, No 2, 69.
95. Panin V.E. and Glikman E.E., Powder metallurgy conference proceedings, Zakopane, 1975, Part 2.
96. Montellbano T., Brett J., Castleman L., Seigle L., Trans. Met. Soc. AIME, 1968, Vol. 242, No 9, 1973.
97. Friedman S., Brett J., Trans. Met. Soc. AIME, 1968, Vol. 242, 2121.
98. Pochivalov Yu.I., Kolobov Yu.R., Korotaev A.D., FMM, 1982, Vol. 54, No.2, 296.
99. Kolobov Yu.R., Korotaev A.D., Kulikov V.A., et al, Materials of the 5th Nat. Conf., Hot Pressing Poweder Metallurgy, Novocherkassk, 1982.
100. Kolobov Yu.R., Korotaev A.D., Kulikov V.A., et al, FMM, 1982, VOl. 54, No.6, 1225.
101. Pochivalov Yu.I., Pinzhin Yu.P., Kolobov Yu.R., Korotaev A.D., Proc. of the 8th Nat. Conf. on Local X-ray Spectral Investigations and their Application, Chernogolovka, 1982.
102. Kolobov Yu.R., Pochivalov Yu.I., Korotaev A.D., Proc. of the 1st Nat. Conf. on the Structure and Properties of Grain Boundaries,Ufa, 1983.
103. Kolobov Yu.R., Pochivalov Yu.I., Pinzhin Yu.P., et al, Dokl. AN SSSR, 1984, Vol. 275, No.3, 616.
104. Kolobov Yu.R., Pochivalov Yu.I., Korotaev A.D., Pinzhin Yu.P., Physics and technology of processing the surface of metals, Leningrad, Nauka, 1984.
105. Kolobov Yu.R., Pochivalov Yu.I., Korotaev A.D., Variation of the structure and properties of alloys based on molybdenum in annealing and creep in the conditions of diffusion contact with nickel, In: Alloys of refractory and rare metals for operation at high temperatures, Moscow, Nauka, 1984.
106. Kolobov Yu.R., Korotaev A.D., Effect of diffusion flows of an impurity on the internal interfaces on the evolution of the substructure and creep of molybdenum, In: Substructural hardening of metals and diffusion methods of investigation, Kiev, Naukova dumka, 1985.
107. Kolobov Yu.R., Ratochka I.V., Marvin V.V., et al, Poverkhnosty. Fizika, Khimiya, Mekhanika, 1984, No.69, 100.
108. Ratochka I.V., Naidenkin E.V., Danilenko V.N., Kolobov Yu.R., FMM, 1995, Vol. 79, No.6, 26.
109. Marvin V.B., Kolobov Yu.R., Izv. Vuzov. Fizika, Tomsk, 1990, Dep. v VINITI 24.09.1990, No.6044-B90.
110. Hongmo Li., Hillert M., Acta met., 1981, Vol. 29, No 12, 1949.
111. Rabkin E., Scripta metall., 1994, Vol.30, 1443.

112. Kajihara M., Gust W., Scripta Mat., 1998, Vol. 38, No 11, 1621.
113. Hillert M., Scripta metall., 1983, Vol.17, No 2, 237.
114. Yamamoto Y., Kajihara M., Acta mater., 1999, Vol. 47, No. 4, 1195.
115. Li C. and Hillert M., Acta metall., 1982, Vol 30, No 6, 1133.
116. Tashiro K., Purdy G.R., Scripta met., 1987, Vol. 21, No 3, 321.
117. Balluffi R.W., Cahn J.W., Acta met., 1981, Vol. 29, No 3, 493.
118. Tashiro K., Purdy G.R., Scripta met., 1983, Vol. 17, No 4, 455.
119. Pan F.D., Balluffi R.W., Acta met., 1982, Vol. 30, No 4, 861.
120. Cahn J.W., Pan I.D., Balluffi R.W., Scripta met., 1979, Vol. 13, No 6, 503.
121. Butrymowiecz D.B., Newbury D.E., Turnbull D., Cahn J.W., Scripta met., 1984, Vol. 18, No 9, 1005.
122. Diffusion and Defect Data. Metals and Alloys (by F.H. Wohlbier), Metals and Alloys, 1980, Vol. 21, 228.
123. Gertsriken S.D., Dekhtyar I.Ya., Diffusion, Moscow, Fizmatgiz, 1960.
124. Den Broeder F.I.A., Klerk M., Vanderberg J.M., Hamm R.A., Acta met.,1983, Vol. 31, No 2, 285.
125. Kim W., Meyrick G., Shewmon P.G., Scripta met., 1983, Vol. 17, No.12, 1435.
126. Den Broeder F.I.A., Nakahara S., Scripta met., 1983, Vol. 17, No.3, 399.
127. Den Broeder F.I.A., Thin. Solid Films, 1985, Vol. 124, No 2, 135.
128. Fedorov G.B. and Smirnov E.A., Metallovedenie i Termich. Obrabotka Metallov, 1981, Vol.15.
129. Parthasavaty T.A., Shawmon P.G., Met. Trans., 1983, Vol. 14, No 7, 2560.
130. Fedorov G.B., Smirnov E.A., Diffusion in reactor materials, Moscow, Atomizdat, 1978.
131. Kolobov Yu.R., Marvin V.B., Ratochka I.V., Korotaev A.D., Dokl. AN SSSR, 1985, Vol. 283, No.3, 605.
132. Kolobov Yu.R., Marvin V.B., Korotaev A.D., Izv. Vuzov. Fizika, 1984, No.12, 97.
133. Kolobov Yu.R., Marvin V.B., FMM, 1989, Vol. 67, No. 6, 1204.
134. Kolobov Yu.R., Marvin V.B., Metallofizika, 1989, Vol. 11, No.4, 3.
135. Glikman E.E., Igoshev V.I., Poverkhnost'. Fizika, Khimiya, Mekhanika., 1984, No.12, 5.
136. Glikman E.E., Igoshev V.I., Zenkova E.K., Nefedov V.G., Poverkhnost'. Fizika, Khimiya, Mekhanika, 1985, No.11, 89.
137. Sarafians N., Z. fur Metallk., 1987, Vol. 78, No. 63, 209.
138. Novikov I.I., Portnoi V.K., Superplasticity of alloys with ultrafine grains, Moscow, Metallurgiya, 1981.
139. Kaibyshev O.A., Superplasticity of industrial alloys, Moscow, Metallurgiya, 1984.
140. Diffusion and defect data (by F.H. Wohlbier), Mater. Rev. Sec., 1976, Vol. 13, No 1-4, 1-292.
141. Ridlin N., Superplasticty of iron-based alloys, In: Superplastic forming of structural alloys, Moscow, Metallurgiya, 1985.
142. D. Liu, W.A. Miller and K.T. Aust., Acta metall., 1989, Vol. 37, No. 12, 3367.
143. Kolobov Yu.R., Proc. of Nat. Conf.: Physics of strength and ductility of metals and alloys, Kuibyshev, 1989.
144. Marvin V.B., Kolobov Yu.R., Poverkhnost'. Fizika, Khimiya, Mekhanika, 1991, No.7, 131.

145. Chan J.W., Balluffi R.W., Scripta met., 1979, Vol. 13, No 6, 499.
146. Kolobov Yu.R., Diffusion-control processes of grain boundaries and plasticity of metallic polycrystals, Novosibirsk, Nauka, Siberian Division, 1998.
147. Klinger L., Rabkin E., Acta mater., 1999, Vol.47, No 3, 725.
148. Konstantinidis D.A. and Aifantis E.C., Scripta Materialia., 1999, Vol.40, No.11, 1235.
149. Klotsman S.M., Uspekhi Fizicheskikh Nauk., 1990, Vol. 160, No.1, 99.
150. Valiev R.Z., Aleksandrov I.V., Nanostructured materials produced by severe plastic deformation, Moscow, Logos, 2000.
151. Mishin Yu., Herzig Ch., Nanostructured materials. Proceedings of the second international conference on nanostructured materials, Stuttgart University, Germany, 1995, Vol.6, .859.
152. Larikov L.N., Metallofixika i Noveishie Tekhnologii, 1995, Vol. 68, No. 2, 99.
153. Keblinski P., Wolf D., Phillpot S.R. and Gleiter H., Scripta Materialia, 1999, Vol. 41, No 6, 631.
154. Klotsman S.M., FMM, 1993, Vol. 76, No.4, 5.
155. Hoefler H.J., Averback R.S., Gleiter H., Phil. Mag. Letts., 1993, Vol.68, No 2, 99.
156. Bokstein B.S., Brose H.D., Trusov L.I., Khvostantseva T.P., Nanostructured materials. Proceedings of the second international conference on nanostructured materials, Stuttart University, Germany, 1995, Vol.6, 873.
157. Kolobov Yu.R., Grabovetskaya G.P., Ivanov M.B., Zhilyaev A.P., Valiev R.Z., Scripta Materialia. In press.
158. Ishida Y., Ichinose H., Kizuka T., Suenaga K., Nanostructured materials. Proceedings of the second international conference on nanostructured materials, Stuttgart University, Germany, 1995, Vol.6, 115.
159. Trampenau J., Bauszus K., Petry W., Herr U., Nanostructured materials. Proceedings of the second international conference on nanostructured materials., Stuttgart University, Germany, 1995, Vol.6, 551.
160. Babanov Yu.A., Blagina L.A., Golovshchikova I.V., et al, FMM, 1997, Vol.83, No.4, 167.
161. Wurschum R., Reimann K., Grub S., Kubler A., Scharwaechter P., Frank W., Kruse O., Carstanjen H.D. and Schaefer H.-E., Phil. Mag. B., 1997, Vol. 76, No. 4, 407.
162. Tanimoto H., Farber P., Wurschum R., Valiev R.Z. and Schaefer H.-E., Nanostructured Materials, 1999, Vol. 12, 681.
163. Tanimoto H., Pasquini L., Prummer R., Kronmuller H. and Schaefer H.-E., Scripta Materialia., 2000, Vol. 42, No 10, 961.
164. Valiev R.Z., Mulyukov R.R., Muulyukov Kh.Ya., et al, Pis'ma v ZhTF, 1989, Vol. 15, No.1, 78.
165. Wurschum R., Kubler A., Gruss S., Scharwaechter P., Frank W., Valiev R.Z., Mulyukov R.R., Schaefer H.-E., Annales de Chimie, 1996, Vol. 21, No. 6-7, 471.
166. Schaefer H.-E., Wurschum R., Gessmann T., Stockl G., Scharwaechter P., Frank W., Valiev R.Z., Fecht H.-J. and Moelle C., Nanostructured materials. Proceedings of the second international conference on nanostructured materials, Stuttgart University, Germany, 1995, Vol.6. 869.
167. Kolobov Yu.R., Grabovetskaya G.P., Ratochka I.V., Kabanova E.V., Naidenkin E.V., Lowe T.C., Annales de Chimie, 1996, Vol.21, No 6-7, 483.
168. Grabovetskaya G.P., Ratochka I.V., Kolobov Yu.R., Puchkareva L.N., FMM,

1997, Vol.83, No.3, 112-115.
169. Kolobov Yu.R., Grabovetskaya G.P., Ratochka I.V., Ivanov K.V., Izv. Vuzov. Fizika, 1998, No.3, 77-82.
170. Kolobov Yu.R., Grabovetskaya G.P., Ivanov M.B., Valiev R.Z., Lowe T.C., Investigations and Applications of Severe Plastic Deformation, Eds. T.C. Lowe and R.Z. Valiev, NATO Science Series, Vol. 80, Dordrecht: Kluwer Academic Publishers, 2000, 261.
171. Ning Wang, Zhirui Wang, Aust K.T. and Erb U., Acta mater., 1997, Vol. 45, No. 4, 1655.
172. Kumpmann A., Guenther B., Kunze H.-D., Mater. Sci. Eng., 1993, Vol.A168, 165.
173. Lian J., Valiev R.Z., Baudelet B., Acta Metall. Mater., 1995, V.43, 4165.
174. Languillaume J., Chmelik F., Kapelski G., Bordeaux F., Nazarov A.A., Canova G., Esling C., Valiev R.Z., Baudelet B., Acta Met.Mater., 1993, V.41, 2953
175. Korznikov A.V., Safarov I.M., Laptionok D.V., Valiev R.Z., Acta metal., mater., 1991, Vol.3, 3193.
176. Korznikov A.V., Ivanisenko Yu.V., Laptionok D.V., Safarov I.M., Pilyugin V.P., Valiev R.Z., Nanostructured Materials, 1994, Vol.4, 159.
177. Korznikov A., Dimitrov O., Quivy A., Korznikova G., Devaud J., Valiev R., Journal de Physique IV. Coll C7, supp. J. de Phys. III, 1995, Vol.5, C7-271.
178. Korznikov A., Dimitrov O., Korznikova G., Annales de Chimie, 1996, Vol.21, No 6-7, 443.
179. Mulyukov Kh.Ya., Korznikova G.F., Abdulov R.Z., Valiev R.Z., Journ. of Magn. and Magn. Mater., 1990, Vol.89, 207.
180. Mulyukov Kh.Ya., Korznikova G.F., Valiev R.Z., Phys. Stat. Sol. (a), 1991, Vol.125, 609.
181. Shen H., Li Z., Guenther B., Korznikov A.V., Valiev R.Z., Nanostructured Materials, 1995, Vol.6, 385.
182. Islamgaliev R.K., Murtazin R.Ya., Syutina L.A., Valiev R.Z., Phys. Stat. Sol. (a), 1992, Vol.129, 231.
183. Gertsman V.Y., Birringer R., Valiev R.Z., Gleiter H., Scr. Met. Mater., 1994, Vol.30, 229.
184. Islamgaliev R.K., Chmelik F., Kuzel R., Mater. Sci. Eng., 1997, Vol.A234-236, 335.
185. Kozlov E.V., Popova N.A., Ivanov Yu.F. et al, Annales de Chimie, 1996, Vol.21, No 6-7, 427.
186. Kolobov Yu.R., et al, Proc. of the 5th Nat. Conf.: Physics and chemistry of ultrafine (nano) systems, Ekaterinburgh, 2000, 245.
187. Research and Application of Metal Matrix Nanocomposites. Report of INTAS project No 1997-1243, 2000.
188. Korznikov A.V., Korznikova G.F., Myshlyaev M.M., et al, FMM, 1997, Vol.84, No.4, 133.
189. Korolev A.V., Deryagin A.I., Zavalishin V.I., Kuznetsov R.I., FMM, 1989, Vol. 68, 672.
190. Mulyukov Kh. Ya., Valeev K.A., Akhmadeev N.A., Nanostructured Materials, 1995, Vol.5, 449.
191. Kolobov Yu.R., Grabovestkaya G.P., Ivanov K.V., Girsova N.V., FMM, 2001, No.5, 105.
192. Koneva N.A., Popova L.N., Ignatenko L.I. at al, In: Investigation and Application of Severe Plastic Deformation. Eds. T.C. Lowe and R.Z. Valiev,

High Technology, 2000, Vol.80, 121.

193. Tumentsev A.N., Tretiak M.V., Korotaev A.D., et al, In: Investigation and Application of Severe Plastic Deformation, Ed. T.C. Lowe and R.Z. Valiev. High Technology, 2000, Vol.80, 127.

194. Tumentsev A.N., Pinzhin Yu.P., Korotaev A.D., et al, FMM, 1998, Vol. 86, No. 6, 110.

195. Korznikov A.V., Idrisova S., Noskova N.I., FMM, 1998, Vol. 85, No.3, 113.

196. Bernshtein M.L., Zaimovskii V.A., Introduction into dislocation theory, Metallurgiya, Moscow, 1968.

197. Zwicker U., Titanium and its alloys, Moscow, Metallurgiay, 1979, 512.

198. Kolobov Yu.R., Kashin O.A., Sagymbaev E.E., et al, Izv. Vuzov. Fizika, 2000, No.1, 77.

199. Valiev R.Z., Kozlov E.V., Ivanov Yu.F., Lian J., Nazarov A.A., Baudelet B., Acta Metall. Mater., 1994, V.42, 2467.

200. Smirnova N.A., Levit V.I., Pilyugin V.I., et al, FMM, 1986, Vol. 62, 566.

201. Gerstman V.Yu., Hoffmann M., Gleiter H., Birringer R., Acta metall. mater., 1994, Vol.42, 3539.

202. Zhang J., Ardell A.J., J. Amer. Ceram. Soc., 1993, Vol.76, 1340.

203. Kochendorfer A., Z. Metallk., 1944, Vol. 105, 193.

204. Aleksandrov I.V., Development and application of methods of x-ray diffraction analysis for examination of the structure and properties of nanostructured materials, Dissertation for the title of the doctor of physico-mathematical sciences, Ufa, 1997.

205. Ahmadeev N.H., Kobelev N.P., Mulyukov R.R., Soifer Ya.M., Valiev R.Z., Acta Metall, Mater., 1993, Vol.41, 1041.

206. Kaibyshev O.A., Valiev R.Z., Grain boundaries and propeties of metals, Moscow, Metalluriya, 1987.

207. Frost G.J. and Ashby M.F., Charts of deformation mechanisms, Chelyabinsk, Metallurgiya, 1989.

208. Ivanisenko Yu.V., Formation of the submicro crystalline structure in iron and steels in severe cold plastic deformation, Dissertation, Ufa, 1997.

209. Musalimov R.Sh., Valiev R.Z., Scripta met., 1992, Vol.27, 1685.

210. Valiev R.Z., Chmelik F., Bordeaux F., Kapelski G., Baudelet B., Scripta metall. mater., 1992, Vol.27, 855.

211. Siegel R.W., In: Proc. NATO ASI "Mechanical properties of ultrafine-grained materials" (Eds. M. Nastasi, D.M. Parkin, H. Gleiter), Dordrecht/Boston/London: Kluwer Head. Publ., 1993, Vol.233, 509.

212. Mishra R.S., Valiev R.Z., McFadden S.X., Mukherjee A.K., Mater. Sci. Eng. A., 1998, Vol.A252, 174.

213. McFadden S.X., Michra R.S., Valiev R.Z., Zhilyaev A.P., Mukherjee A.K., Nature, 1999, Vol.398, 684.

214. Senkov O.N., Froes F.H., Stolyarov V.V., Valiev R.Z., Liu J., Scr. Mater., 1998, Vol.38, 1511.

215. Portnoi K.I., Babich B.N., Dispersion-hardened materials, Moscow, Metallurgiya, 1974.

216. Single crystal fibres and materials reinforced with them, Edited by R.L. Albert, Moscow, Mir, 1973.

217. Daneliya E.P., Rozenherg V.M., Internally oxidised alloys, Moscow, Metallurgiya, 1978.

218. Paninn V.E., Dudarev E.F., Bushnev L.S., Structure and mechanical properties of substitutional solid solutions, Moscow, Metallurgiya, 1971.

219. Schilling W.F., Grant N.S., Powder Metallurgy International. 1973, Vol.5, No 3. 117.
220. Smirnyagin A.P., Smirnyagina N.A., Belova A.V., Industrial and non-ferrous metals and alloys, Moscow, Metallurgiya, 1974.
221. Sauer C., Weißgärber T., Dehm G., et al, Zeitschrift für Metallkunde, 1998, Vol.89, No 2, 119.
222. Sekino T., Niihara K., Metal nanocomposites. Nanostructured Materials, 1995, Vol.6, 665.
223. Grundmann U., Gerner M., Heilmaier M., et al, Mat. Sci. and Eng., 1997, Vol. A.234-236, 505.
224. Naser J., Ferkel H., Riehemann W., Mat. Sci. and Eng., 1997, Vol. A.234-236, 470.
225. Cai B., Kong Q.P., Lu L., Lu K., Scripta Mater., 1999, Vol.41, No.7, 775.
226. Synk J.E., Vedula K., Material Science and Technology, 1987, Vol.3, 72.
227. Naser J., Riehemann W., Ferkel H., Mat. Sci. and Eng., 1997, Vol. A.234-236, 467.
228. Buchgraber W., Islamgaliev R.K., Kolobov Yu.R., Amirkhanov N.M., In: Investigation and Application of Severe Plastic Deformation. Ed. T.C. Lowe and R.Z. Valiev. High Technology, 2000, Vol.80, 267.
229. Kolobov Yu.R., Ivanov K.V., Grabovetskaya G.P., In: Investigation and Applications of Severe Plastic Deformation. Ed. T.C. Lowe and R.Z. Valiev. High Technology, 2000, Vol.80, 339.
230. Alexandrov I.V., Zhu Y., Lowe T., et. al, Metallurgical and Materials Transactions, 1998, Vol. A29, 2253.
231. Islamgaliev R.K., Akhmadeev N.A., Mulyukov R.R., Valiev R.Z., Metallofizika, 1990, No.12, 31.
232. Valiev R.Z., Islamgaliev R.K., Kuzmina N.F., Li Y., Langdon T.G., Scripta Mater., 1999, Vol.40, 117.
233. Mishra R.S., Valiev R.Z., McFadden S.X., Islamgaliev R.K., Mukherjee A.K., Phil.Mag. In print.
234. Kulikov V.A., Zaitseva N.M., Kolobov Yu.R., Bushnev L.S., Structure and properties of dispersion-hardened alloys, Tomsk, Tomsk State University, 1998.
235. Amirkhanov N.M., Islamgaliev R.K., Valiev R.Z., FMM, 1998, Vol. 86, No.3, 99.
236. Musalimov R.Sh., Valiev R.Z., FMM, 1992, No.2, 95.
237. Abdulov R.Z., Valiev R.Z., Karsil'nikov N.A., Electron microscopy and strength of materials, Kiev, Institute of Powder Metallurgy, 1989.
238. Panin V.E., In: Investigation and Applications of Severe Plastic Deformation. Ed. T.C. Lowe and R.Z. Valiev. High Technology, 2000, Vol.80, 203.
239. Panin V.E., Deerevyagina L.S., Valiev R.Z., Fiz. Mezomekhanika, 1999, Vol. 2, No.1-2, 89.
240. Kolobov Yu.R., Ivanov K.V., Grabovetskaya G.P., Islamgaliev R.K., Structure and properties of nanocrystalline materials, Ekaterinburg, 1999.
241. Dudarev E.F., Bakach G.P., Grabovetskaya G.P., Chernova L.V., Fiz. Mezomehanika, 2001, Vol. 4, No.1, 97-104.
242. Honeycombe R.W., Plastic deformation of metals, Moscow, Mir, 1972.
243. Kaibyshev O.A., Plasticity and superplasticity of metals, Moscow, Metallurgiya, 1975.
244. Sanders P.G., Yongdall C.J., Weertman J.R., Mat. Science and Engineering. 1997, Vol.A234-236, 77.

245. Tyumentsev A.N., Panin V.E., Derevyagina L.S., et al, Fiz. Mezomekahnika, 1999, Vol.2, No.6, 115.
246. Valiev R.Z., Islamgaliev R.K., FMM, 1998, Vol.85, No.3, 161.
247. Ivanov K.V., Kolobov Yu.R., Grabovestkaya G.P, et al, Proc. of the 5th Intl. Conf.: Evolution of defective structure in condensed media, Barnaul, 2000.
248. Movchan B.A., Problemy Prochnosti, 1999, No.1, 3.
249. Grabovetskaya G.P., Ratochka I.V., Kabonova E.V., et al, Izv. Vuzov. Fizika, 1994, No.12, 947.
250. Ivanov K.V., Ratochka I.V., Kolobov Yu.R., Nanostructured Materials, 1999, Vol.12, 947.
251. Mishra R.S., In.: Ultrafine Gained Materials, Ed. Mishra R.S., Semiating S.L., Lowe T.C., et al, Minerals, Metals and Materials Sosiety,- 2000, 421.
252. Mishra R.S., Valiev R.Z., Mukkerjii A.K., Mater. Sci. Eng. A., (In print).
253. Mishra R.S. Mc Fadden S.X., Valiev R.Z., Mukkerjii A.K., J. Metals, 1999, Vol.51, No 1, 37.
254. Kovalevski Z., Arch. met., 1992, Vol.37, .65.
255. Wang N., Wang Z., Aust K.T., Erb U., Acta Metal Mater., 1995, Vol.43, No 2, 519.
256. Wang D.L., Kong O.P., Shui J.P., Scripta Metal Mater., 1994, Vol.31, No 1, 47.
257. Weertman J.R., Mater. Sci. Eng., 1993, A166, 161.
258. Hayes R., Tllkamp V., Lavernia E., Scripta Mater., 1999, Vol.41, No7, 743.
259. Rittner M.N., Weertman J.R., Eastman J.A., et al, Mat. Sci. Eng., 1997, Vol. A237, 185.
260. Chokshi A.H., Rosen A., Karch J., Gleiter H., Scripta Metall., 1989, Vol.23, 1679.
261. Cobl R.L., J.Appl. Phys., 1963, Vol.34, 1679.
262. Horvath J., Berringer R., Gleiter H., Sol. Stat. Commun. - 1987. - Vol.62. - P.319.
263. Sherby O.D., Wadsworth J., Progress in Materials Science, 1989, Vol.33, 169.
264. Valiev R.Z., Rasumovskii I.M., Sergeev V.I., Phys. stat sol. – 1993. – Vol.139. – P. 321.
265. Gleiter H., Chalmers B., High-angle grain boundaries, Mir, Moscow 1975.
266. Kolobov Yu.R., Grabovetskaya G.P., Ratochka I.V., Ivanov K.V., Nanostructured materials, 1999, Vol.12, 1127.
267. Garofalo F., Laws of creep and long term strength of metals and alloys, Moscow, Metallurgiya, 1968.
268. Naidenkin E.V., Grabovetskaya G.P., Kolobov Yu.R., Ratochka I.V., FMM, 1999, Vol. 88, No.4, 101.
269. Kolobov Yu.R., Ratochka I.V., FMM, 1990, No.8, 185.
270. Valiev R.Z., Kaibyshev O.A., Acta metall., 1983, Vol.31, 2121
271. Grabskii M.V., Structural superplasticity of metals, Moscow, Metallurgiya, 1975.
272. Islamgaliev R.K., Buchgraber W., Kolobov Yu.R., et al, Proceedings of ICSMA-12. California. (In print).
273. Shankar M.L., In: Investigation and Application of Severe Plastic Deformation. Eds. T.C. Lowe and R.Z. Valiev. High Technology, 2000, Vol.80, 273.
274. Rösler J., Arzt E., Acta Met., 1990, Vol.38, 671.

275. Rösler J., Joos R., Arzt E., Metall. Trans., 1992, Vol.A23A, 1521.
276. Kolobov Yu.R., Ratochka I.V., J. Mater Sci and Technol., 1995, Vol.11, No.1, 38.
277. Grabovetskaya G.P., Kolobov Yu.R., Zverev I.K., Izv. Vuzov: Fizika, 1992, No.7, 16.
278. Valiev R.Z., Xhabubulin V.G., FMM, 1992, No.7, 16.
279. Ratochka I.V., Kolobov Yu.R., Metallofizika, 1991, No.7, 40.
280. Grabovetskaya G.P., Kolobov Yu.R., Marvin V.B., Prokof'ev S.I., FMM, 1990, No.9, 193.
281. Grabovetskaya G.P., Naidenkin E.V., Kolobov Yu.R., Ratochka I.V., Izv. Vuzov. Fizika, 1997, No.7, 119-125.
282. Kaibyshev O.A., Superplasticity of Alloys, Intermetallides and Ceramics, Berlin: Springer, 1992.
283. Ball A., Hutchison M.M., Met. Sci. J., 1969, Vol.3, No.1, 1.
284. Glikman E.E., Igoshev V.I., Poverkhnost'. Fizika, Khimiya, Mekhanika, 1989, No.3, 104.
285. Zhang K., Aleksandrov I.V., Kilmametov A.R., at al, Phys. D.: Appl. Phys., 1997, Vol.30, 3008.
286. Geguzin Ya.E., Kibets V.I., Chekanov M.I., In: Proc. of the Nat. Conf.: Physics of the strength and plasticity of metals and alloys, Kuibyshev 1985.
287. Raj R., Achby M.F., Met. Trans., 1972, Vol.3, No.7, 1937.
288. Shtremel' M.A., Strength of alloys, Part 1: Lattice defects: a textbook for higher technical schools, Moscow, Moscow Institute of Steel and Alloys, 1999.
289. Bogachev I.N., Weinstein A.A., Volkov S.D., Introduction into statistical metal science, Moscow, Metallurgiya, 1972.
290. Shermergor T.D., Theory of elasticity of microheterogeneous media, Moscow, Nauka, 1977.
291. Dudarev E.F., Microplastic deformation and yield limit of polycrystals, Tomsk, Tomsk State Technical University, 1988.
292. Microplasticity, Translated from the English, Edited by V.N. Geminova and A.G. Rakhshtadt, Moscow, Metallurgiya, 1972.
293. Friedel G., Dislocations, Moscow, Mir, 1967.
294. Rakhshtadt A.G., Spring steels and alloys, Moscow, Metallurgiya, 1971.
295. Dudarev E.F., Kashin O.A., Kolobov Yu.R., et al, Izv. Vuzov. Fizika, 1998, No.12, 20.
296. Ivanova V.S., Terent'ev V.F., Nature of the fatigue of metals, Moscow, Metallurgiya, 1975.
297. Troshchenko V.T., Deformation and failure of metals in multicycle loading, Kiev, Naukova dumka, 1981.
298. Kocanda S., Fatigue failure of metals, Moscow, Metallurgiya, 1975.
299. Dudarev E.F., Pochivalova G.P., Nikitina N.V., Izv. Vuzov. Fizika, 1990, No.3, 29.
300. Structural materials, a handbook, edited by B.N. Arzamasov, Moscow, Mashinostroenie, 1990.
301. Kolachev V.A., Physical metallurgy of titanium, Moscow, Metallurgiya, 1976.
302. Kolachev V.A., Mal'kov A.V., Physical fundamentals of failure of titanium, Moscow, Mashinostroenie, 1987.
303. Modification and alloying of the surface by laser, iron and electron beams, Edited by G. Poute, Moscow, Mashinostroenie, 1987.
304. Sulima A.M., Evstigneev M.I., Quality of the surface layer and fatigue strength

of components of creep-resisting and titanium alloys, Leningrad, Mashinostroenie, 1974.

305. Kovalenko V.S., Verkhoturov A.D., Golovko L.F., Podchernyaeva I.A., Laser and electroerosion hardening of metals, Moscow, Nauka, 1986.

306. Valiev R.Z. Structure and mechanical properties of ultrafine-grained metals., Mater. Sci. A. Eng., 1997, Vol.A234-236, 59.

307. Stolyarov V.V., Alexandrov I.V., Kolobov Yu.R., at al, Fatigue-99. Proc. of the 7-th Int. Fatigue Congress. China, 1999, Vol.3-4, 1435.

308. Kolobov Yu.R., Kashin O.A., Dudarev E.F., et al, Current problems of modern material science, Proceedings, Baikal'sk, 1999.

309. Abramov O.V., Khorbenko I.G., Shvegla Sh., Ultrasound treatment of materials, Moscow, Mashinostroenie, 1984.

310. Rakhimyanov Kh.M., Iskhakova G.A., Elektronnaya Obrabotka Materialov., 1990, No.5, 9.

311. Iskhakova G.A., Rakhimyanov Kh.M., Elektronnaya Obrabotka Materialov., 1990, No.5, 22.

312. Abramov V.O., Abramov O.V., Gradov O.M., et al, Materialovedenie, 1997, No.809, 7.

313. Panin V.E., Klimenov V.A., Bezborodov V.P., et al, Fizika i Khimiya Obrabotki Materialov, 1993, No.6, 77.

314. Perevalova O.B., Kornienko L.A., Bezborodov V.P., et al, Fizika i Khimiya Obrabotki Materialov, 1997, No.3, 82.

315. Klimenov V.A., Ivanov Yu.F., Kolomeets N.P., Shepel V.M., Nechoroshkov O.N., In: Energy amd Environmental Aspects of Tribology, 5th International Symposium INSYCONT98, Cracov, Poland, 1998, 83.

316. Postnikov V.S., Internal friction in metals, Moscow, Metallurgiya, 1974.

317. Krishtall M.A., Golovin S.A., Internal friction and structural metals, Moscow, Metallurgiya, 1976.

318. Blanter M.S., Golovin I.S., Golovin S.A., et al, Mechanical spectroscopy of metallic materials, Moscow, Publishing House of International Engineering Academy, 1994.

319. Dudarev E.F., Kashin O.A., Kolobov Yu.R., Pochivalova G.P., Izv. Vuzov. Fizika, 2001.

320. Investigations and applications of severe plastic deformation. Eds. T.C. Lowe, R.Z. Valiev, Dordrecht, Kluwer Acad. Publ., 2000, 394.

321 Nieh T.G., Wadsworth J., Sherby O.D., Superplasticity in metals and ceramics, Cambridge Univ. Press, 1994, Vol. 6, 164.

322. Mukherjee A.K., Treatise on materials science and technology, Ed. R. Arsenault, New York, Academic Press, 1994, Vol. 6, 164.

323. Tikhonov A.S., Effect of superplasticity of the metals and alloys, Moscow, Nauka, 1978.

324. Griffiths P., Hammond C., Acta Met., 1972, Vol.20, No.7, 935.

325. Likhachev V.A., Myshlyaev M.M., Olevskii S.S., Chuchman T.N., Acta Met., 1974, Vol.22, No.7, 829.

326. Lukhachev V.A., Myshlyaev M.M., Sen'kov O.N., Kuz'mina S.L., FMM, 1979, Vol. 47, No.6, 1271.

327. Kuznetsova R.I., Zhukov N.N., FMM, 1977, Vol.44, No.6, 1277.

328. Zhukov N.N., Kuznetsova R.I., Poida V.P., FMM, 1979, Vol. 48, No.6, 1282.

329. Kuznetsova R.I., Zhukov N.N., Kaibyshev O.A., Valiev R.Z., Phys. Stat. Sol. A., 1982, Vol. 70, No.2, 371.

330. Maevskii I.A., Processing of creep-resisting alloys, Mashinostroenie, Moscow,

1964.
331. Naziri H., Pearce R., Scr. Met., 1969, Vol.3, No.11, 807.
332. Naziri H., Pearce R., J. Inst. Met., 1970, Vol. 98, No.3, 71.
333. Kaibyshev O.A., Kazachkov I.V., Salikhov S.Ya., Aca Met., 1978, No. 3, 124.
334. Kaibyshev O.A., Kazachkov I.V., Salikhov S.Ya., Acta Met., 1978, Vol. 26, No.12, 1887.
335. Dunlop G.L., Reid J.D., Taplin D.M., R. Met. Trans., 1971, Vol.2, No.8, 2308.
336. Portnoi V.K., Vovikov I.I., Efremov B.N., Kozhanov V.A., Izv. Vuzov. Tsv. Metallurgiya, 1979, No.5, 103.
337. Frommeyer G., Z. Metallik., 1979, Vol.70, No.9, 573.
338. Paton N.E., Hamilton C.H., Met. Trans., 1979, Vol.10A, No.2, 241.
339. Valiev R.Z., Vergazov A.N., Gertsman V.Yu., Crystal-geometrical analysis of intercrystalline boundaries, Moscow, Nauka, 1991.
340. Sutton A.P., Balluffi R.W., Interfaces in crystalline materials, Oxford: Clarendon Press, 1995.
341. Valiev R.Z., Mater. Sci. Forum., 1997, Vol. 243-245, 207.
342. Valiev R.Z., Islamgaliev R.K., Stolyarov V.V., et al, Mater. Sci. Forum., 1998, Vol. 269-272, 969.
343. Horita Z., Smith D.J., Nemomoto M., et al, J. Mater. Res., 1998, Vol. 13, 446.
344. Furukawa M., Horita Z., Nemomoto M., et al, J. Mater. Res., 1996, Vol. 11, 2128.
345. Valiev R.Z., Musalimov R.Sh., FMM, 1994, Vol. 78, No.6, 114.
346. Valiev R.Z., Gayanov R.M., Yang H.S., Mukherjee A.K., Scr. Met. Mater., 1991, Vol. 25, 1945.
347. McFadden S.X., Valiev R.Z., Mukherjee A.K., Superplasticity in nanocyrstalline Ni_3Al, 2001.
348. Valiev R.Z., Song C., McFadden S.X., et al, Phil. Mag. A., 2001, Vol. 81, 25.
349. Mishra R.S., Mukherjee A.K., Superplasticity and superplastic forming, Ed. Ghosh A.K., Bieler T.R., San Antonio, 1998.
350. Valiev R.Z., Mukherjee A.K., Scripta Material, 2001, Vol. 44, No.809, 1747.
351. Kaibyshev O.A., Markelov A.A., Izv. Vuzov. Chern. Metallurgiya, 1973, Vol. 7, 119.
352. McFadden S.X., Zhilyaev A.P., Mishra R.S., Mukherjee A.K., Mater. Let., 2000, Vol.45, No.6, 345.
353. Valiev R.Z., Salimonenko D.A., Tsenev N.K., et al, Scripta Mater., 1997, Vol. 37, 724.
354. Komura S., Berbon P.B., Furukawa M., et al, Scripta Mater., 1998, Vol. 38, 1851.
355. Berbon P.B., Furukawa M., Horita Z., et al, Phil. Mag. Let., 1998, Vol. 78, No.4, 313.
356. Valiev R.Z., Islamgaliev R.K., Mater. Sci. Forum, 1999, Vol. 304-306, 39.
357. Mishra R., Valiev R.Z., Mukherjee A.K., Nanostructured Mater., 1997, Vol.2, 430.
358. Valiev R.Z., Gertsman V.Yu., Kaibyshev O.A., Phys. Stat. Sol., 1986, Vol. 97, 11.
359. Everhart L., Engineering properties of nickel and nickel alloys, New York, Plenum Press, 1971, 20.

Index